Springer Tracts in Civil Engineering

Springer Tracts in Civil Engineering (STCE) publishes the latest developments in Civil Engineering - quickly, informally and in top quality. The series scope includes monographs, professional books, graduate textbooks and edited volumes, as well as outstanding PhD theses. Its goal is to cover all the main branches of civil engineering, both theoretical and applied, including:

- Construction and Structural Mechanics
- Building Materials
- Concrete, Steel and Timber Structures
- Geotechnical Engineering
- Earthquake Engineering
- Coastal Engineering; Ocean and Offshore Engineering
- Hydraulics, Hydrology and Water Resources Engineering
- Environmental Engineering and Sustainability
- Structural Health and Monitoring
- Surveying and Geographical Information Systems
- Heating, Ventilation and Air Conditioning (HVAC)
- Transportation and Traffic
- Risk Analysis
- Safety and Security

Indexed by Scopus

To submit a proposal or request further information, please contact:
Pierpaolo Riva at Pierpaolo.Riva@springer.com (Europe and Americas) Wayne Hu at wayne.hu@springer.com (China)

More information about this series at https://link.springer.com/bookseries/15088

Kewei Liu · Cristoforo Demartino · Zhi Li ·
Qinghui Liu · Yan Xiao
Editors

2019 International Bamboo Construction Competition

From the Concepts to the Realized Pavilions

INBAR
INTERNATIONAL BAMBOO
AND RATTAN ORGANISATION

Springer

Editors
Kewei Liu
International Bamboo and Rattan Organisation
Beijing, China

Zhi Li
College of Civil Engineering and Architecture
Zhejiang University
Hangzhou, Zhejiang, China

Yan Xiao
Zhejiang University
Haining, Zhejiang, China

Cristoforo Demartino
Zhejiang University
Haining, Zhejiang, China

Qinghui Liu
International Bamboo and Rattan Organisation
Beijing, China

ISSN 2366-259X ISSN 2366-2603 (electronic)
Springer Tracts in Civil Engineering
ISBN 978-3-030-91989-4 ISBN 978-3-030-91990-0 (eBook)
https://doi.org/10.1007/978-3-030-91990-0

This Springer imprint is published by the registered company Springer Nature Switzerland AG
The registered company address is: Gewerbestrasse 11, 6330 Cham, Switzerland

Foreword by Ali Mchumo

Bamboo is one of the fastest growing species on earth, and woody bamboos have been used in buildings for thousands of years. It provides millions of households across the world with employment and building materials. It is often used as a fast-growing alternative to timber or other building materials, and its products can have a low or even negative carbon footprint across their lifecycle. Additionally, well-designed bamboo structures are known to withstand earthquakes and tropical storms remarkably well.

INBAR decided to organize 2019 International Bamboo Construction Competition (IBCC2019), because in recent years, the global modern bamboo construction sector has made remarkable achievements, but the capacity building of bamboo construction professionals lags behind the resources' potential. As an intergovernmental organization with 48 Member States, INBAR has been promoting bamboo as an environmentally friendly substitute for other building materials in the past 20 years. Our aim is to provide sustainable nature-based solutions to housing shortages, climate change mitigation and adaptation, and livelihood development. Due to lack of capacity among bamboo construction professionals, bamboo's potential in the construction sector has not yet been fully realized. INBAR is trying to work with our partners to redress the gap. It is for this reason that INBAR jointly launched this competition with other 40 global institutions in the beginning of 2019 in order to promote the awareness of global universities to train students to use bamboo as a construction material.

IBCC2019 has been the first major global competition on bamboo construction in the past ten years, which is aimed at university students of architecture, civil engineering, landscape, and other construction-related courses from across the world to explore the potential construction applications for both bamboo poles and engineered bamboo products.

IBCC2019 has achieved great success and attracted attentions of students from around 30 countries. Students from various universities in the world actively participated and brought their creative and high-quality works, which laid a solid foundation for the success of the competition. Many universities even officially announced IBCC2019 as a kind of short-term course, which fully reflects the importance that universities attach to it. After the mid-term evaluation by 23 global senior bamboo experts, four finalists from Belgium, Ethiopia, and China got the chance to demonstrate their ideas into real bamboo houses in the INBAR Garden, and another 11 also were invited to present their designs to on-site jurors in INBAR's "Bamboo Eye" pavilion. Sixteen international awards were selected in total. During the 2019 Beijing International Horticultural Expo, thousands of visitors came to the INBAR Garden enjoyed those designs by global students who presented the future of bamboo construction sector.

The book records the designs concept, implementation process, and personal experience of students from Asia, Africa, Oceania, America, and Europe who participated in this competition. These vivid memories are the most precious things for the competition which are what the organizer wants to record mostly. I sincerely hope it will be helpful for encouraging more students and faculties to pay attention to bamboo in the future.

Finally, on behalf of INBAR, I would like to express my gratitude to Zhejiang University and Springer Nature for their support in the publication of this very interesting book which I am sure readers will enjoy.

Dar es Salaam, Tanzania
September 2021

Ali Mchumo
Director General of INBAR

Foreword by David Trujillo

Bamboo provides millions of families across the world with employment and construction materials. Its products can have a low or even negative carbon footprint across their lifecycle. Bamboo reduces pressure on forest resources and can replace concrete-block or clay-brick housing. Well-designed bamboo structures are known to withstand earthquakes and tropical storms remarkably well. Paradoxically, the teaching of bamboo construction is rarely integrated into university curricula across the world.

The first International Bamboo Construction Competition sought to address this imbalance by inviting students of construction-related courses from across the world to design and build a bamboo structure within the framework of the 2019 Beijing Horticultural Expo. INBAR felt that the Horticultural Expo was an excellent opportunity to raise public awareness about bamboo's potential in construction, and the IBCC2019 was a cornerstone of this initiative.

The theme of the competition was inspired by the Horticultural Expo and required entrants to design the prototype for a multifunctional garden building. The brief required that the building should be able to operate as an office, warehouse, educational space, or relaxation suite. An important consideration for sustainable construction is not only to use sustainable materials, but to ensure that comfort can be achieved in the least energy-intensive way. Therefore, the indoor temperature of the garden building would have to remain pleasant throughout Beijing's summer without resort to mechanical ventilation or artificial cooling. Therefore, entrants would have to consider passive forms of cooling.

Due to the nature of the location and event, the building was necessarily temporary. This did not imply though, that it should be designed to have a short life span. Indeed, the building needed to be easily demountable and transportable, so that it could be displayed elsewhere, after the Horticultural Expo had ended. With this in mind, and also with the aim to educate future designers about the need to design bamboo structures to be durable, the brief required them to limit the use of details that would hinder the durability of bamboo.

Bamboo is a very versatile material; therefore, it was a reasonable expectation that the building would be made mostly from bamboo poles and/or engineered bamboo products. Nevertheless, bamboo does have its limitations. A good designer knows how to combine materials to obtain an optimal result. Therefore, participants could use a limited amount of other materials, but only wherever the aim was to enable or enhance bamboo's potential, or wherever clearly bamboo was inappropriate. This way, young designers would explore bamboo's potential, but learn to accept the material's limitations.

As is true for all building projects, the building is needed to be built on budget and on time. With this requirement, the brief ensured participants gave due regard to buildability and cost. Ultimately, a good design is not that which is achieved at a disproportionate price.

These may have been 9 m^2 garden buildings designed by university students, but they contain the aesthetic grace, sustainability considerations, structural engineering, and construction management of a full-scale building. The IBCC not only achieved the goal of promoting bamboo among budding professionals from across the world, but also demonstrated that it is an inspirational and richly educational resource. Hopefully, it will be emulated across the world.

September 2021

David Trujillo
Chair of INBAR Bamboo Construction
Task Force; Assistant Professor, School
of Energy, Construction and
Environment
Coventry University
Coventry, UK

INBAR and IBCC 2019

INBAR

Established in 1997, the International Bamboo and Rattan Organisation (INBAR) is an intergovernmental development organization that promotes environmentally sustainable development using bamboo and rattan. It is currently made up of 48 Member States. In addition to its Secretariat Headquarters in China, INBAR has five regional offices in Cameroon, Ecuador, Ethiopia, Ghana, and India.

Bamboo, the fast-growing grass plant, and rattan, the spiky climbing palm, can be important nature-based solutions to a number of pressing global challenges, for poverty alleviation, green trade, climate change mitigation and adaptation, resilient construction, and environmental protection. The mission of INBAR is to improve the well-being of producers and users of bamboo and rattan within the context of a sustainable bamboo and rattan resource base, by consolidating, coordinating, and supporting strategic and adaptive research and development.

INBAR's work is based around the following strategic goals:

1. Promoting bamboo and rattan in socioeconomic and environmental development policies at national, regional, and international levels;
2. Coordinating inputs on bamboo and rattan from a growing global network of members and partners and representing the needs of members on the global stage;
3. Sharing knowledge and communicating lessons learned, providing training, and raising awareness of the relevance of bamboo and rattan as plants and commodities;
4. Fostering adaptive research and on-the-ground innovation by promoting pilot case studies and supporting upscaling of best practices across INBAR Member States.

INBAR has been making a real difference to the lives of millions of people and environments around the world for over 20 years. Its achievements include raising standards; promoting safe, resilient bamboo construction; restoring degraded land;

building capacity for bamboo and rattan practitioners; creating hundreds of thousands of jobs; and informing green policy and Sustainable Development Goal (SDG) objectives, particularly, ending poverty (SDG 1), providing a source of clean and renewable energy (SDG 7), building affordable housing (SDG 11), encouraging sustainable consumption and production (SDG 12), combatting climate change (SDG 13), protecting life on land (SDG 15), and creating partnerships for the achievement of the Goals (SDG 17).

IBCC 2019

The International Bamboo Construction Competition 2019 (IBCC 2019) is an international competition opened in Beijing, China, from March 1 to July 19, 2019. It was initiated by INBAR and its partners in 2019, aiming at encouraging global university students of architecture, civil engineering, landscape, and other related majors to explore the potential of bamboo as a construction material.

The 2019 Beijing Horticultural Expo was held in Beijing, China. At this event, INBAR had a 3600 m^2 garden, which houses its own 1600 m^2 pavilion. The pavilion was built from round bamboo poles and engineered bamboo products. INBAR believed that this was an excellent opportunity to raise public awareness about bamboo's potential in construction.

Therefore, "A Garden Pavilion" was given as the theme of IBCC 2019. Approximately 500 students from 30 countries participated the competition and submitted 70 designs in total. Four of 15 finalists realized their design at INBAR Garden, and others presented their design concept to the jury. After the competition, all the shortlisted teams were invited to write a scientific article to introduce their design concept and experience.

IBCC 2019 has four key outputs as follows: (1) improved students' understanding of bamboo as a construction material and the whole construction system; (2) stimulated students' creative inspiration; (3) provided a fair opportunity to global students; and (4) trained students to showcase their ideas in a scientific way.

Contents

About the Editors

Mrs. Liu Kewei is the Coordinator of Global Bamboo Construction Programme at the International Bamboo and Rattan Organisation (INBAR). She graduated from Beijing Jiaotong University with master degree in Structural Engineering in 2007 and is now a Ph.D. candidate of School of Civil Engineering at Tsinghua University. Since 2008, she has been working on promoting bamboo construction across the world. She managed around 20 international bamboo construction projects across Asia, America, Africa and Europe. She is managing the INBAR Bamboo Construction Task Force, comprising more than 30 bamboo construction experts from around 20 countries. As Convenor of Working Group 12 on Structural Use of Bamboo within the Timber Structures Technical Committee of International Organization for Standardization (ISO TC165 WG12), she is leading the development and revision of international standards for bamboo construction. Her academic achievements include more than 20 peer-reviewed papers, 5 technical standards, two monographs as the first author, one academic conference proceedings in English as one editor-in-chief, the Chinese-language version of an English book as the first translator.

Dr. Cristoforo Demartino is an Assistant Professor (2019-) at the Zhejiang University/University of Illinois at Urbana-Champaign Institute (ZJUI) at the Zhejiang University International Campus in Haining, China. He is also Adjunct Assistant Professor (2020-) at the University of Illinois at Urbana-Champaign (UIUC), Urbana-Champaign, Illinois, U.S. B.Sc. (2008) and M.Sc. (2010) full marks with honours in Civil Engineering at the Mediterranea University of Reggio Calabria. 2nd level University Master degree at University of Rome "La Sapienza" with a thesis on "Evaluation of the seismic economic losses of precast concrete industrial buildings" in 2012. Ph.D. in Structural Engineering at the University of Naples "Federico II" with a thesis on "Aerodynamics and aeroelastic behavior of ice-accreted bridge cables" in 2014. Adjunct Professor of a course on RC structures at Second University of Napoli for the period 2014-2015. Post-Doc position at the University of Rome "La Sapienza" for "Regional seismic risk of industrial facilities including uncertainties and economic implications" for the period 2015-2016. Post-Doc of the College of Civil Engineering at the Nanjing Tech University for a project named "Structural and non-structural elements subjected to extreme loads" for the period 2016-2018. He is the author of more than 100 publications in international journals (46) and conferences (57) mainly on the topics of Structural Engineering, Wind Engineering, Earthquake Engineering, and Structural Dynamics. Cristoforo Demartino has given over 30 invited lectures and he is an active reviewer in more than 50 international journals. He is PI or co-PI in different research projects.

Dr. Zhi Li is an Associate Research Professor in the College of Civil engineering and Architecture, Zhejiang University. From Feb. 2015 to June 2016, he worked as a postdoctoral researcher at Sapienza University of Rome. From June 2016 to August 2020, he was an Assistant Professor at Nanjing Tech University. His main research interests are in modern bamboo & timber structures, bio-composite materials, and its probability characteristics. He has authored and co-authored more than 20 international journal papers related to these topics. He is also serving as PI and co-PI in research projects funded by NSF China, and China MOST.

Mrs. Qinghui Liu is the Programme Officer of Global Programme at the International Bamboo and Rattan Organisation (INBAR). Mrs. Liu took part in organising several high-level international events on bamboo and rattan since 2017. She was one of main members of the Organizing Committee of Global Bamboo and Rattan Congress (BARC 2018) in 2018. She also took part in organising a series of activities held at INBAR Pavilion of the International Horticultural Exhibition, Beijing, China in 2019 (Beijing Expo 2019), during which she contributed to the International Bamboo Construction Competition (IBCC 2019). In recent two years, Mrs. Liu also starts to participate in the management of several bamboo and rattan projects in Africa.

Dr. Yan Xiao is a Distinguished Professor, Program Director for Energy, Environment and Infrastructure Sciences, in the Zhejiang University – University of Illinois at Urbana Champaign Institute (ZJUI), Zhejiang University International Campus in Haining, China. His academic career spanned in Japan, US and China. He serves as the associate editor for the ASCE Journal of Structural Engineering, Journal of Bridge Engineering, and editorial board member of the Journal of Constructional Steel Research. He is elected fellows of the American Society of Civil Engineers (ASCE) and the American Concrete Institute (ACI). Passionate about bamboo, Dr. Xiao focuses and publishes pioneered research on engineered bamboo and holds many patents and the trademark technology GluBam®. In 2008, he was awarded the Best of What's New in 2008 by the Popular Science Magazine, and was named as the Popular Science Innovator of 2008. Prof. Xiao's team has designed and built many demonstration bamboo-based houses and bridges in China and Africa.

The 2019 International Bamboo Construction Competition

Kewei Liu, Cristoforo Demartino, Zhi Li, Qinghui Liu, and Yan Xiao

Abstract In 2019, the International Bamboo and Rattan Organisation (INBAR) launched the International Bamboo Construction Competition 2019 (IBCC 2019). IBCC 2019 is an international contest aimed at university students of architecture, civil engineering, landscape, and other construction-related courses. The objective of the competition is to invite students to explore the potential construction applications for both bamboo poles and engineered bamboo products. IBCC 2019 was sponsored by several important companies working in the development of innovative bamboo products for the construction and design industries. The competition was organized on three different stages: pre-selection and qualification; selection of the three projects to be realized; and realization and award of the best project. This chapter opens with a discussion on the nature of the competition and then summarizes the key steps of the IBCC 2019 from the ideation to the award ceremony. The key design goals are widely discussed together with a deep analysis of the potentiality of bamboo for the construction industry.

Keywords IBCC 2019 · Student competition · Bamboo · Temporary structures · Design projects

K. Liu · Q. Liu
International Bamboo and Rattan Organisation, Beijing, China

C. Demartino (✉) · Y. Xiao
Zhejiang University-University of Illinois at Urbana Champaign Institute, Zhejiang University, Hangzhou, Zhejiang, PR China
e-mail: cristoforodemartino@intl.zju.edu.cn

Department of Civil and Environmental Engineering, University of Illinois at Urbana-Champaign, Urbana, IL 61801, USA

Z. Li
College of Civil Engineering and Architecture, Zhejiang University, Hangzhou 310058, Zhejiang, PR China

K. Liu
School of Civil Engineering, Tsinghua University, Beijing, China

K. Liu et al. (eds.), *2019 International Bamboo Construction Competition*, Springer Tracts in Civil Engineering, https://doi.org/10.1007/978-3-030-91990-0_1

1 Introduction

Competitions play a role in motivating students to perform and excel and offer a lot more rewards than just the winning prize [1]. Competitions offer a chance for participants to gain substantial experience, showcase skills, analyze and evaluate outcomes and uncover personal aptitude. Competitions also encourage students to adopt innovative techniques and develop their ideas and skills.

Learning by doing refers to a theory of education expounded by American philosopher John Dewey [2]. It's a hands-on approach to learning, meaning students must interact with their environment to adapt and learn. Dewey implemented this idea by setting up the University of Chicago Laboratory School. His views have been important in establishing practices of progressive education. By letting structural engineering and architecture students explore questions using architectural design methods, they creatively and systematically addressed holistic questions while maintaining a technical depth. The approach may serve as a model to increase engineering students' ability to insightfully contribute to solutions for complex societal problems. One way of invigorating engineering and architecture education is to combine engineering with liberal arts, grounding students more firmly in creative and innovative design, social responsibility, and critical thinking. These three outcomes are key to forming our future architects and civil engineers.

The environmental impact of the construction industry contributes to global warming [3]. Construction projects emit large amounts of carbon dioxide and methane. Infrastructure developments cause pollution and produce waste. The construction sector accounts for 36% of worldwide energy demand, and 40% of the global energy and process-related emissions [4]. As the output of the construction industry multiplies, so can its damaging effects. Therefore, a reduction of the environmental impact of the construction industry is necessary. As a matter of fact, sustainability is no more an option but a way of living that we need to rapidly adopt for our survival and sustainability of humanity. In this context, China announced during the 75th session of the UN General Assembly (UNGA 75) will aim to have CO_2 emissions peak before 2030 and achieve carbon neutrality before 2060.

The United Nations Sustainable Development Goals (SDGs) adopted by 193 countries at the Paris Accord (COP21) encompass the pressing issues of sustainability, to be addressed holistically on the social, environmental and economical fronts [4]. SDG 11 involves making human settlements sustainable, safe, and resilient. Again, the designers of MEP systems have an important part to play, especially where they can minimize water and energy use. Construction material with better thermal properties can further reduce HVAC plant capacity, with up to 10% reduction achieved due to the use of material with a better heat transfer coefficient. There are numerous ways to decrease the negative environmental impact on your construction project. By utilizing green and reusable technology, keeping project length to a minimum, limiting fuel use, along with other suggestions listed, you are sure to make a difference in the impact on the environment. Thus, it is essential to understand the emission reduction potential of new technological options in the construction industry.

Bamboo is a truly remarkable green building material [5]. Bamboo is a rapidly renewable material that has many applications in construction. Bamboos are a diverse group of evergreen perennial flowering plants in the subfamily *Bambusoideae* of the grass family *Poaceae* [6]. The origin of the word "bamboo" is uncertain, but it probably comes from the Dutch or Portuguese language, which originally borrowed it from Malay or Kannada. Bamboos are of notable economic and cultural significance in South Asia, Southeast Asia, and East Asia, being used for building materials, as a food source, and as a versatile raw product. Bamboo, like wood, is a natural composite material with a high strength-to-weight ratio useful for structures. Bamboo is twice as strong as concrete and slightly stronger than steel. Bamboo has an unparalleled growth rate in the plant world. The world record for the fastest growing plant belongs to certain species of the 45 genera of bamboo, which have been found to grow at up to 91 cm per day [7]. Because the regeneration rate for bamboo forests is much quicker than any other woody plant, it is a far superior sustainable choice. The main drawback in using bamboo is that being a natural material presents large irregularities therefore it is generally difficult to exactly determine its properties (geometrical and mechanical). To overcome this issue, different researchers proposed engineered bamboo products. Engineered bamboo products result from processing the raw bamboo culm into a laminated composite, similar to glue-laminated timber products [8]. These products allow the material to be used in standardized sections and have less inherent variability than the natural material [9].

In 2019, the International Bamboo and Rattan Organisation (INBAR) launched the International Bamboo Construction Competition 2019 (IBCC 2019). IBCC 2019 was launched with the aim to combine all the elements summarized before: competition for students, learning by doing, reduction of the environmental impact of the construction industry, and use of bamboo. This chapter opens with a discussion on the nature of the competition and then summarizes the key steps of the IBCC 2019 from the ideation to the award ceremony.

2 Genesis of the Competition

In 2019, the International Bamboo and Rattan Organisation (INBAR) launched the IBCC 2019. The logo of IBCC 2019 was designed by the students of the organizing committee, see Fig. 1. The logo is inspired by the following elements: international

Fig. 1 Logo of IBCC 2019 (designed by Ke Ma)

by color, bamboo by shape, construction by the view of perspective, competition by the shape of small arrows.

The genesis of this student competition is oriented to the idea that student competitions can inspire architecture, civil engineering, landscape, and other construction-related students to greater achievement. By adding an element of collaboration to the competition (teamwork), the organizers ensured that every student has a chance to participate and achieve the same level of learning as the high-achieving students. The competition aims at improving the idea of environmentally-friendly buildings: structures that are resource-efficient throughout their life-cycle, throughout the processes of design, construction, operation, maintenance, renovation, and eventual demolition. Environmentally-friendly buildings represent a paradigm shift in the field of architecture and a more sustainable way to secure the future of urbanization.

To this aim, the opportunities offered by the use of bamboo (raw or engineered) for construction are explored by IBCC 2019. Bamboo is the fastest-growing and most versatile plant on Earth. For centuries, bamboo has played an indispensable part in the daily life of millions of people in tropical countries. In the last decades, it has gained increasing importance as a substitute for timber. In recent years, the global modern bamboo construction sector has made remarkable achievements: hundreds of large commercial projects are built with bamboo. In November 2018, a quick-assembly bamboo slum housing design won the prestigious "Cities for our Future Challenge" competition, coordinated by the Royal Institute of Chartered Surveyors. However, the capacity building of bamboo construction professionals' lags behind the resources' potential.

IBCC 2019 was established to redress this gap. IBCC 2019 was an international contest aimed at university students. The objective of the competition is to invite students to explore the potential construction applications for both bamboo poles and engineered bamboo products. The general goals of IBCC2019 and expected outcomes are summarized in Fig. 2. IBCC 2019 started with the aim to combine all the elements summarized before: competition for students, learning by doing, reduction of the environmental impact of the construction industry, and use of bamboo.

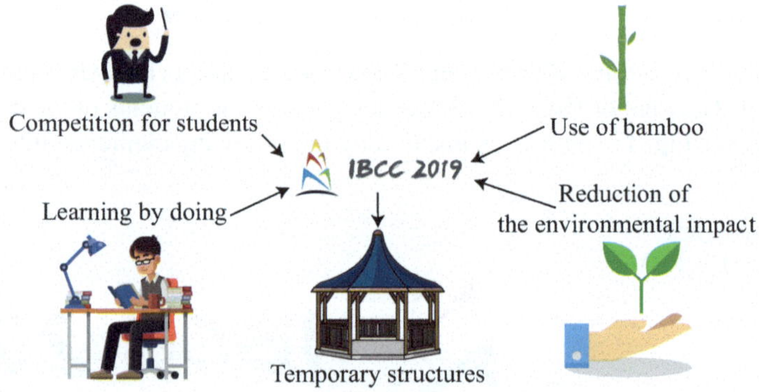

Fig. 2 General goals of the competition and expected outcomes

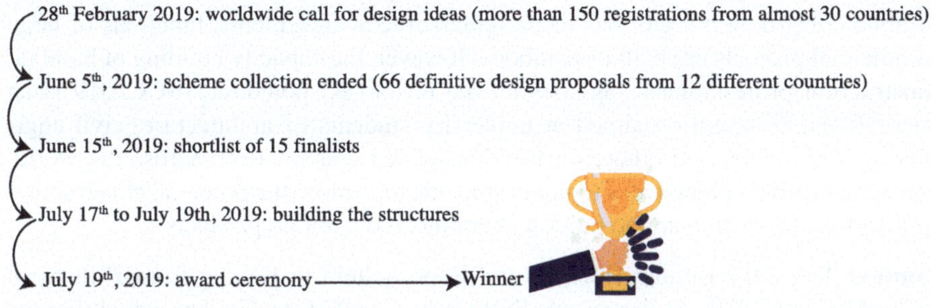

Fig. 3 Timeline of the IBCC2019

The design competition asked to design a temporary structure. These structures are erected to fill a temporary need, lasting for hours, days, weeks, and sometimes months instead of years. A typical temporary structure is designed to provide shelter to people, products, equipment, animals, or anything else which needs to be sheltered from the elements. In particular, IBCC2019 required to design a prototype of a temporary structure for a multi-functional garden building able to operate as a temporary office, warehouse, educational space, or coffee shop with a construction area of 9 m^2. The building is to be made mostly from bamboo poles and/or engineered bamboo products and be characterized by a low environmental impact. Additionally, the building should be easily demountable and transportable, and though it is a temporary structure, detailing known to limit the durability of bamboo should be avoided.

The competition was organized in four stages (see Fig. 3). In the first stage, a worldwide call for design ideas (Sect. 3.1). In the second stage, a jury of 23 members selected a shortlist of 15 finalists (Sect. 3.1). Finally, the top three finalists built the real bamboo house in the INBAR Garden during the 2019 Beijing Horticultural Expo (Sect. 4). At the end of the construction process, an award ceremony was organized at the 2019 Beijing Horticultural Expo (Sect. 5).

3 The Competition: From Registration to the Shortlist

3.1 The Worldwide Call for Design Ideas

The competition was launched on the 28th of February 2019. In the following, the worldwide call for design ideas is reported.

Background Bamboo provides millions of households across the world with employment and building materials. It is often used as a fast-growing alternative to timber or other building materials, and its products can have a low or even negative carbon footprint across their lifecycle. Well-designed bamboo structures are known to withstand earthquakes and tropical storms remarkably well. In recent years, the global modern bamboo construction sector has made remarkable achievements: hundreds of large commercial projects are built of bamboo. However, the capacity building of bamboo construction professionals' lags behind the resources' potential. IBCC2019 is an international competition aimed at university students of architecture, civil engineering, landscape, and other construction-related courses from across the world. The competition's objective is to invite students to explore the potential construction applications for both bamboo poles and engineered bamboo products.

Context The 2019 Beijing Horticultural Expo is held in Beijing from 29th April to 7th October 2019. At this event, INBAR has a 3000 m^2 Garden, which houses our own 1600 m^2 Pavilion. The pavilion was built from round bamboo poles and engineered bamboo products (Simi et al. A Review on Round Bamboo Structural Applications and Perspectives).

Who Can Participate in the Competition? BCC2019 is open to university students from the departments of architecture, civil engineering, landscape, or other construction-related courses. Full-time on-campus students, includes undergraduate students, master, and Ph.D. candidates, can participate in the competition and should enter as a team. Teams are constituted by one advisor and three students. There should be at least one student from architecture, landscape architecture, or architectural technology, and one student from civil or structural engineering. It is the responsibility of the advisor to demonstrate the structural safety of the design, particularly for any proposal that is to be built in the INBAR Garden of the 2019 Beijing Horticultural Expo.

Topic of the Competition IBCC 2019 is inspired by the 2019 Horticultural Expo and requires entrants to design the prototype for a multi-functional garden building. The design competition asked to design a temporary structure. These structures are erected to fill a temporary need, lasting for hours, days, weeks, and sometimes months instead of years. A typical temporary structure is designed to provide shelter to people, products, equipment, animals, or anything else which needs to be sheltered from the elements. In particular, IBCC required to design a prototype of a temporary structure for a multi-functional garden building able to operate as a temporary office, warehouse, educational space, or coffee shop with a construction area of 9 m^2. The building is to be made mostly from bamboo poles and/or engineered bamboo products and be characterized by a low environmental impact. Additionally, the building should be easily demountable and transportable, and though it is a temporary structure, detailing known to limit the durability of bamboo should be avoided. The following technical characteristics were required:

- A construction area of 9 m^2 (3 m $\times$ 3 m), as well as a projection area no larger than 16 m^2 (4 m $\times$ 4 m), including cantilever structural elements such as canopy. Plat concrete foundation is available for each project.
- The building should be one or two stories. The maximum height of the first floor is 3 m and the total is 5 m. The first floor should be a closed space (also a part is fine) while the second (optional) open. The first floor must have the minimum requirement of one door and one window.

Table 1 Design strength values of engineered bamboo along the bamboo fiber direction in MPa

Bending strength	Compression strength	Tension strength	Shear strength along bamboo strips	Elastic modulus
20	17	8	4	9500

- The indoor temperature should remain pleasant throughout Beijing's summer (around 24–26 °C) without resort to mechanical ventilation or artificial cooling.
- The building should be easily demountable and transportable, and though it is a temporary structure, detailing known to limit the durability of bamboo should be avoided.
- The building is to be made mostly from bamboo poles and/or engineered bamboo products. A limited amount of other non-structural materials may be used, but only with the aim of enabling or enhancing bamboo's potential, or where clearly bamboo is inappropriate. It should be assumed that only Moso bamboo (Phyllostachys edulis) is available for bamboo poles unless otherwise agreed with the organizing committee. Engineered bamboo products available will be announced by the organizing committee. The design strength values of engineered bamboo can be estimated as in Table 1 (if no data from manufacturers is available).
- The total budget for all building materials is USD 10,000, and should not be exceeded. It must be buildable by a team of four people over three days (three 8 h periods).

Requirements of Submission The submission was required to have graphic boards and a descriptive report. In particular, the following specifications were provided to the different teams:

- Graphic boards: 2 graphic boards in vertical on a rigid support, format UNI-EN-ISO A1 containing: plans, profiles, and sections representing the architectural design in scale 1:50; simulations, representations, sketches, profiles, or perspective views (with an appropriate scale) that would be needed to understand the design proposal. Perspective views useful for understanding the project (simulations, photomontage); drawings of constructive details useful for understanding the architectural elements of the used materials and constructive technology, scale 1:10. Processing files must be printed and should not exceed the size of 50 MB for the single file.
- The Descriptive Report: maximum 6 pages plus the cover—A4 format—printed single side on white paper—body character ARIAL 11 pt. Single line—maximum 5000 characters—with pictures or explanatory drawings. The report may contain images and diagrams of the proposal. The report shall illustrate the guiding principles of the choices and description of the design solutions.

On June 5th, 2019, at the end of the worldwide call for design ideas, the IBCC2019 received more than 150 registrations from almost 30 countries. After a preliminary selection of compliance with the rules of the context, 66 design ideas from 12 countries we accepted to be judged by the jury for to be possibly included in the shortlist of 15 finalists.

3.2 *The Shortlist of 15 Finalists*

The Jury Committee of IBCC2019 was composed of 23 jurors from the top universities and leading companies in the world in the area of bamboo. The 23 jurors are from different countries so that the equity and transparency of the competition are guaranteed.

Chairman of the Jury Committee is Kai Cui (Fig. 4). Cui Kai is Chief Architect of China Architecture Design and Research Group, Professor of Tianjin University, Practical Professor of Tsinghua University, Professor of University of Chinese Academy of Sciences, and editorial board member of several architectural magazines. He is also the designer of the China Pavilion in the 2019 Beijing Horticultural Expo.

The other 22 jurors of the committee are:

Denamo Addissie, Assistant Professor of Addis Ababa University, Ethiopia.
Yann Olivier Barnet, Director of Institute of Housing, Urban Planning and Construction of the University San Martín, Peru.
Edwin Zea Escamilla, Head of Sustainable Real Estate at the Centre for Corporate Responsibility and Sustainability of the University of Zürich, Switzerland.
Romildo D. Toledo Filho, Professor of Federal University of Rio de Janeiro, Brazil.

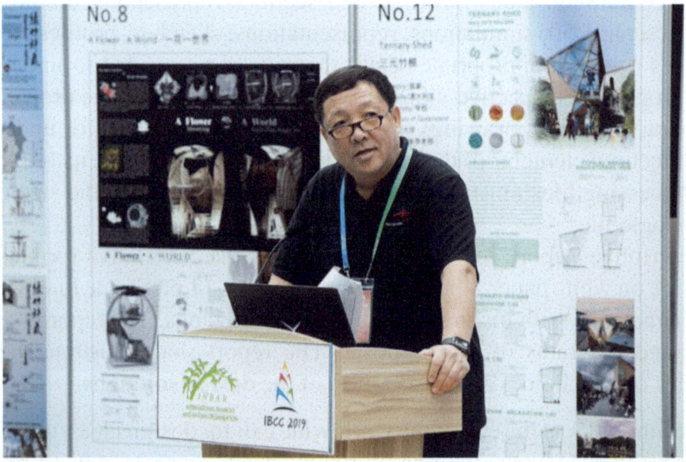

Fig. 4 Kai Cui (Chairman of the Jury Committee) during the IBCC2019 award ceremony

Kent Harries, Professor of the University of Pittsburgh, USA.
Sebastian Kaminski, Senior Structural Engineer of Arup, UK.
Sanjeev Kappe, Founder Director of Konkan Bamboo & Cane Development Center, India.
Emmanuel Appiah Kubi, Research Scientist of CSIR-Forestry Research Institute, Ghana.
Gisela Loehlein, Head of Architecture Department Xi'an Jiaotong-Liverpool University, China.
Liu Kewei, Global Bamboo Construction Programme Coordinator of INBAR, China.
Mauricio Cardenas, CEO of Studio Cardenas Conscious, Italy.
Haitao Li, Professor of Nanjing Forestry University, China.
Luis Felipe Lopez, Head of Product Development at Base Bahay Foundation, the Philippines.
Martin Tam, Chair of Federation of Hong Kong Industries Bamboo Industry Committee, China.
David Trujillo, Chair of INBAR Construction Task Force, Assistant Professor of Coventry University, UK.
Yuanfeng Wang, Director of Sustainable Civil Engineering Committee of the Chinese Society for Urban Studies, Professor of Beijing Jiaotong University, China.
Yan Xiao, Professor of Zhejiang University, China.
Qingfeng Xu, Deputy Chief Engineer of Shanghai Research Institute of Building Sciences, China.
Jun Yang, Professor of Tsinghua University, China.
Li Zhang, Chief Editor of World Architecture, Professor of Tsinghua University, China.
Chen Zhao, Professor of Nanjing University, China.
Xi Zheng, Professor of Beijing Forestry University, China.

After several rounds of selection, the jury selected 15 finalists from the total 66 design projects received. The official announcement was given on June 15th, 2019. The 15 finalists are listed in Table 2.

The top three finalist teams in in Table 2 were invited to build the real bamboo house in the INBAR Garden during the 2019 Beijing Horticultural Expo. The description of the projects reported in Table 2 is provided in this book in the next chapters.

4 The Top Three Finalists: Building the Structures

The top three finalist teams were invited to the 2019 Beijing Horticultural Expo to build the proposed design project from July 17th to July 19th, 2019. The three teams were from Europe, Africa, and China. The three finalists proposed quite different

Table 2 Fifteen finalists selected from the jury

Name	University	Advisor	Team 1	Team 2	Team 3
House Loti	Université de Liège ULiege	Vincent Denoël	Audrey Mertens	Fantine Fontaine	Chen Qi
In-Box House	Addis Ababa University	Boja Muluneh	Lelissa Erkissa	Mikyas Tekle	Simon Mucheye
Machine of bamboo	Nanjing University	Xiao Fu	Huimin Wu	Xiaonan Li	Sicheng Zhou
Crouching · hidden	Beijing Jiaotong University	Zeng Zhongzhong	Cai Yueqian	Qian Xiaowen	Liu Dingyi
Nomad	University of Chinese Academy of Sciences	Ren Congcong	Yang Jindi	Jin Liu	Yan Xin
Bloom	Zhejiang university	Zhuo Xin	Yang Chenchen	Rui Guanbao	Wang Chidi
Confronting the performance of bamboo in architecture	Ethiopian Institute of Architecture, Building Construction and City Development	Alemayehu Sargo Dalibiso	Ermias Fentaw Marye	Beka Debela Chali	Yeneneh Tefera Negash
A Flower A World	North China University of Technology	Zhao Xiangdong	Zhao Sijia	Zhang Hongmei	Zhang Ziwei
Bamboo field	Xiamen University	Lin Yuxin	Zhang Yin	Qin Ruonan	Gong Qing
Push and pull varying tranquility bamboo tectonic	Xiamen University	Yuxin Lin	Sihan Wang	Yu Wang	Hao Qiu
Bending Blossom	Xi'an Jiaotong Liverpool University	Thomas Alois Wortmann	Zhao Zhang	Henry Pantin	Caipeng Yan
Ternary shed	The University of Queensland	Kimberly Baber	Mateo Gutierrez Gonzalez	Harry Frazer Mills	Hamish Buchhorn
The bud pavilion	Xiamen University	Lin Yuxin	Urali Halimati	Guo Yanfang	Hong Yangyi
Bamboo's verve	Chongqing Jiaotong University	Biao Huang	Xinrui Wei	Jiping Ma	Junfeng Xiong
Name	Xi'an Jiaotong-Liverpool University	David Vardy	Shi Luhang	Gao Xinxin	Xie Haitian

design ideas. The first two projects were based on raw bamboo (House Loti and In-Box House, see Table 2) while the last one was on engineered bamboo (Machine of Bamboo, see Table 2). Moreover, the project Bloom (see Table 2) realized the structure although not selected in the top three finalist teams.

A collection of photos of the three days of construction at the 2019 Beijing Horticultural Expo to build the proposed design project from July 17th to July 19th, 2019, is reported in Fig. 5. The students helped by some local workers worked for three full days to build the structures. At their arrival, they found only ready the foundations of the structures made of reinforced concrete. The final structures realized are shown in Fig. 6. These structures are described in Mertens et al. House Loti; Erkissa et al. In-Box House; Wu et al. Machine of Bamboo.

House Loti combines Western and Eastern construction, the traditional and the modern, to create a pavilion building that is adaptive, sustainable, and pleasant. The word "blossom" inspired a dynamic design, being able to actually bloom: petals composing the walls of the closed flower can blossom into sunshade panels and seats.

Fig. 5 Some photos of the three days of construction at the 2019 Beijing Horticultural Expo to build the proposed design project from July 17th to July 19th, 2019

Fig. 6 Top three realized finalist projects. (Left: House Loti from the Université de Liège; center: In-Box House from Addis Ababa University; right: Machine of Bamboo from Nanjing University)

In-Box House is a multifunctional bamboo garden house designed to be constructed in a combination of bamboo poles with prefabricated bamboo lumber. The prefabricated modular bamboo lumbers with simple connection details make it easy to be constructed on-site. The prefab modular boxes are inserted in the bamboo structural frames in many functional layouts according to the functional purposes.

Machine of Bamboo originates from the desire of relaxing in nature. This building has two stories. The construction is covered by a type of semi-transparent material, sunshine board, which is not only easy to install but also keeps privacy and creates good daylighting. Engineered bamboo is used as the main structural material. The building is easily demountable, transportable, and reusable. Even though it is a temporary structure, the long-term durability of structural members has been guaranteed.

5 The Award Ceremony

On July 19th, 2019, the award ceremony hold place at the INBAR Garden Pavilion—Bamboo Eye (see Fig. 7). During the event, three finalist projects presented the structures to the jurors. Moreover, the remaining selected 15 finalists presented their design ideas.

The jurors selected the best projects in the following order:

1. House Loti, designed by a team from Université de Liège. (Team: Audrey Mertens, Fantine Fontaine, Chen Qi; Advisor: Vincent Denoël)
2. Machine of Bamboo, designed by a team from Nanjing University. (Team: Huimin Wu, Xiaonan Li, Sicheng Zhou; Advisor: Xiao Fu)

Fig. 7 Group photo during the award ceremony on July 19th, 2019 in front of the Inbar Garden Pavilion—Bamboo Eye

3. In-Box House, designed by a team from Addis Ababa University. (Team: Lelissa Erkissa, Mikyas Tekle, Simon Mucheye; Advisor: Boja Muluneh)

6 Conclusions and Future Developments

IBCC 2019 demonstrated the potentiality of the use of bamboo in constructions. IBCC 2019 received more than 150 registrations from almost 30 countries, and 66 final design ideas from 12 countries. A jury with jurors from top universities and leading companies around the world selected a shortlist of fifteen finalists and three projects to be realized. The three projects were successfully designed and showed high quality; the next chapters will describe them in a detailed way.

As a suistanable and fast-grown resource, with superior physical and mechanical properties, bamboo offers great potential as an alternative to wood. The bamboo-based industry has vast potential for generating income and employment, especially in rural areas. Therefore, future editions of this context should deeply explore the potentiality of bamboo in constructions.

Acknowledgements IBCC 2019 was sponsored by Ganzhou Sentai Bamboo Wood Company of China, Hongyazhuyuan Science and Technology Company of China, Zhujing Bamboo Company of China and Beijing Sino-bamboo Culture and Technology Limited Company.

References

1. Verhoeff, T. 1997. The role of competitions in education. *Future world: Educating for the 21st Century*, 1–10.
2. Thompson, P. 2010. Learning by doing. *Handbook of the Economics of Innovation* 1: 429–476.
3. Lin, B., and H. Liu. 2015. CO_2 mitigation potential in China's building construction industry: A comparison of energy performance. *Building and Environment* 94: 239–251.
4. Bebbington, J., and J. Unerman. 2018. Achieving the United Nations sustainable development goals. Accounting, Auditing & Accountability Journal.
5. Wang, J.S., C. Demartino, Y. Xiao, and Y.Y. Li. 2018. Thermal insulation performance of bamboo-and wood-based shear walls in light-frame buildings. *Energy and Buildings* 168: 167–179.
6. Liese, W., and M. Köhl. 2015. *Bamboo-the plant and its uses*. Tropical Forestry: University of Hamburg.
7. Guinness World Records webpage. https://www.guinnessworldrecords.com/world-records/fastest-growing-plant/. Last accessed 26 Apr 2021.
8. Sharma, B., A. Gatóo, M. Bock, and M. Ramage. 2015. Engineered bamboo for structural applications. *Construction and Building Materials* 81: 66–73.
9. Xiao, Y., Q. Zhou, and B. Shan. 2010. Design and construction of modern bamboo bridges. *Journal of Bridge Engineering* 15 (5): 533–541.

Use of Bamboo in Constructions

Zhi Li, Kewei Liu, Cristoforo Demartino, and Yan Xiao

Abstract The application of bamboo in construction is briefly introduced in this chapter. The basic mechanical properties of bamboo under different loading directions are given. The compression and tensile stress–strain models in the bamboo fiber direction are reported. In terms of round bamboo, engineered bamboo, and hybrid bamboo structures, bamboo structures are introduced with various construction examples. The carbon emission of bamboo construction was also discussed, compared with other mainstream construction forms. These research results and construction practices validate that bamboo is a promising natural source that can be used as a sustainable construction material.

Keywords Bamboo structure · Bamboo construction · Round bamboo · Engineered bamboo

Z. Li (✉)
College of Civil Engineering and Architecture, Zhejiang University, Hangzhou 310058, China
e-mail: li_zhi@zju.edu.cn

K. Liu
International Bamboo and Rattan Organization, Beijing 100102, China

C. Demartino · Y. Xiao
Zhejiang University-University of Illinois Institute (ZJUI), Zhejiang University, Haining 314400, China

C. Demartino
Department of Civil and Environmental Engineering, University of Illinois at Urbana-Champaign, Urbana, IL 61801, USA

Y. Xiao
Sonny Astani Dept. of Civil and Environmental Engineering, Univ. of Southern California, Los Angeles, CA 90089, USA

K. Liu
School of Civil Engineering, Tsinghua University, Beijing, China

K. Liu et al. (eds.), *2019 International Bamboo Construction Competition*, Springer Tracts in Civil Engineering, https://doi.org/10.1007/978-3-030-91990-0_2

1 Bamboo as a Construction Material

The use of bamboo in construction is prevalent in many bamboo-rich regions and has been practiced for centuries. The coordinate system of different bamboo products used for construction is given in Fig. 1. Various studies [1, 2] indicate that bamboo is one of the most effective natural materials that can be used as a sustainable construction material with high performance in terms of high strength-to-density ratio, good thermal insulation performance, and easy-to-manufacture. Such material has been successfully used for modern residential buildings, bridges, wind blades, and urban water pipelines in recent twenty years worldwide.

The nature of it dominates the mechanical properties of bamboo as a unidirectional fiber-reinforced composite material. Researches [3–7] have been conducted on different bamboo products' strength and modulus properties under different loading directions. According to the authors' recent study [7], Moso bamboo strips' average tension strength along its fiber direction (1-direction) is about 130 MPa, with a stand deviation (St. D) of 16 MPa. The average compressive strength of Moso bamboo strips along fiber direction (1-direction) is about 57 MPa, with St. D of 5 MPa. The modulus value is around 10 GPa. Strength values perpendicular to the fiber direction (2-direction) are about 6 MPa (St. D 1 MPa) for tension and 14 MPa (St. D 1 MPa) for compression. As shown in Fig. 2, the simplified segmental linear models can be derived for all bamboo products, as given in Fig. 1. As illustrated in this figure, seven parameters are used to describe the model, including values of f_{t1}, E_{t1}, f_{c1}, and $E_{c1,}$ as given above. ε_{tu} is the ultimate tension strain, 12,000 με for round bamboo, bamboo strips, and unidirectional bamboo panels. It is 7800 με for multidirectional laminated bamboo panels. ε_{c0} is the compression strain at maximum stress, which is 23255 με for unidirectional bamboo products, and 10,000 με for multidirectional

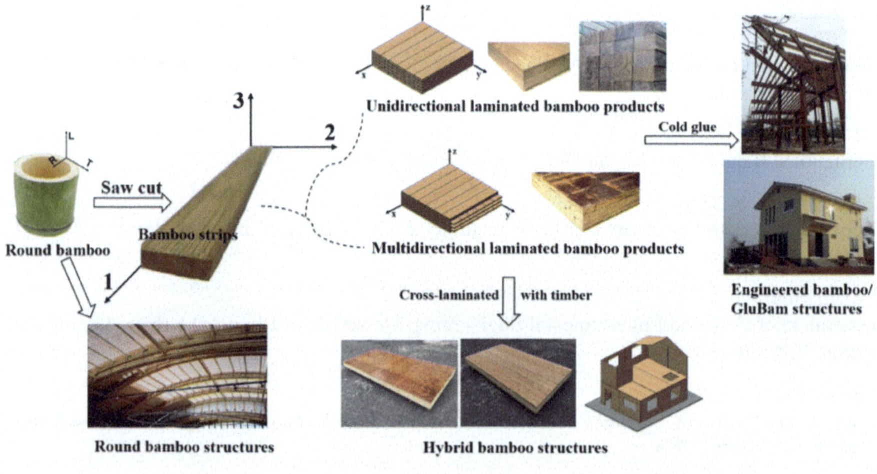

Fig. 1 The coordinate system of different bamboo products and bamboo structures

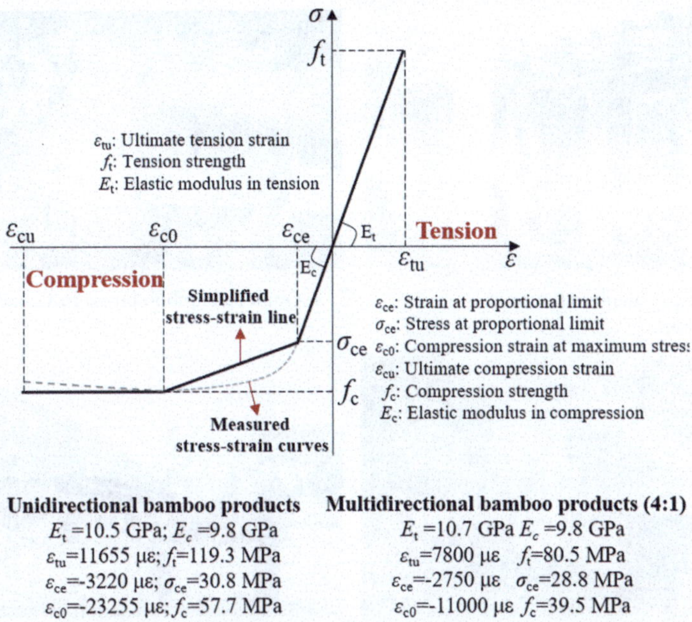

Fig. 2 Stress–strain model of bamboo products

bamboo panels. The ultimate compressive strain ε_{cu} is about 30,000–35,000 $\mu\varepsilon$ for all bamboo products. Parameter σ_{ce} is the measured stress at the proportional limit state, 30.8 and 28.8 MPa for thick and multidirectional bamboo. Strain ε_{ce} corresponding to the proportional limit stress is 3220 and 2750 $\mu\varepsilon$ for unidirectional and multidirectional bamboo products, respectively. The performance of bamboo structures with stresses on their main-fiber direction can be estimated with this stress–strain model.

Specifically, for bamboo used in residential construction, three types of structural forms are widely used nowadays: round bamboo structures, engineered bamboo structures, and hybrid bamboo structures with conventional construction materials as concrete, steel, and timber.

2 Round Bamboo Structures

Bamboo has been indispensable in human life, and its history being used as building and bridge structural materials can be traced back as long as wood. Its natural properties are pretty comparable to those of wood. Today, people in many countries still keep the habits of living in traditional round bamboo buildings, such as Bahareque bamboo houses in Colombia (Fig. 3a), Quincha bamboo houses in Ecuador (Fig. 3b), Sidama bamboo houses in Ethiopia (Fig. 3c) and Gan-Lan bamboo houses in China (Fig. 3d) [8]. Most traditional round bamboo houses were designed and constructed without

a. Bahareque bamboo house in Colombia

b. Quincha bamboo house in Ecuador

c. Sidama bamboo house in Ethiopia

d. Ganlan bamboo house in China

Fig. 3 Traditional round bamboo houses around the world (Photo source: Fig. 3d provided by Huang Wenkun, others provided by INBAR) [8]

the participation of architects and engineers. Local people built them according to the experience passed down from their ancestors [8].

However, due to the irregular geometrical shape and anisotropic material properties, the bamboo culms are challenging to be widely used as an industrial building material in modern construction. It is urgent that the people need modern construction technologies to consolidate the inheritance of traditional skills. Since the 1970s, modern round bamboo structures with architects' participation in design and construction have developed unprecedentedly. A group of excellent architects from Colombia solved the bearing capacity problem of hollow round bamboo joints by pouring cement mortar into bamboo poles and anchoring them with metal connections. It makes bamboo used to build larger public buildings (Fig. 4) and brings a new chapter in developing modern round bamboo structures. In Asia, where it is suitable for bamboo growth, many excellent round bamboo structures (Fig. 5) also have emerged in the past two decades.

In some cases, modern prefabricated technologies are also used for the construction of round bamboo structures. One trial project was the application of round bamboo culms with engineered bamboo panels in shear walls. Due to prefabrication and assembly construction requirements, the walls must be made into standard units and then transported to the construction site to form the structure through simple connectors. The outer framing is made of GluBam and then nailed with the

a. A teaching building in Colombia. (photo by Wang Ge)

b. Jenny Garzon Bridge in Colombia. (photo by Liu Kewei)

c. A Hotel in Panama. (Photo by Simón Veléz Architects)

d. Nomadic Art Museum in Mexico. (Photo by Simón Veléz Architects)

Fig. 4 Modern round bamboo structures in Latin American countries [8]

sheathing panels. The round bamboo columns as inner studs are connected with the sheathing through U-hoop connectors and connected to the framing through L-shape connectors. A demonstration house using this type of bamboo shear walls was built in Hunan, as shown in Fig. 6. With the promotion of modern technology, the round bamboo structure has promising development potential.

3 Engineered Bamboo Structures

In recent 20 years, engineered bamboo products have gradually gained commercial importance. In many countries, engineered bamboo products are used in commercial projects for decoration purposes, like indoor or outdoor flooring, wallboard, ceiling, etc. However, engineered bamboo is mainly used in China for its structural uses, where some local pioneer architects started to apply them in different projects.

As shown in Fig. 7a, architect Li Daode created a youth hostel through digital design method and generation logic in Niubeishan Mountain in Sichuan Province. The rolling roof of the hostel is echoing with the mountains in front of it and clouds behind it, utilizing bamboo scrimber made from *Dendrocalamus affinis*, which is a new engineered bamboo material that emerged in the Chinese market. Architect Cao

a. OBI Great Hall in Java, Indonesia. (photo by Andry Widyowijatnoko)

b. Naman Retreat in Vietnam. (photo by Cai Wei)

c. Eco-Children's Activities and Education Center, Thailand.

d. INBAR Pavilion at 2019 Beijing International Horticultural Expo, China. (photo by Wang Xudong)

Fig. 5 Modern round bamboo structures in Asian countries. The project of Fig. 5c won the Second Prize of the International Bamboo and Rattan Products Design Competition organized by INBAR at the 2010 Shanghai World Expo, provided by Olav Bruin [8]

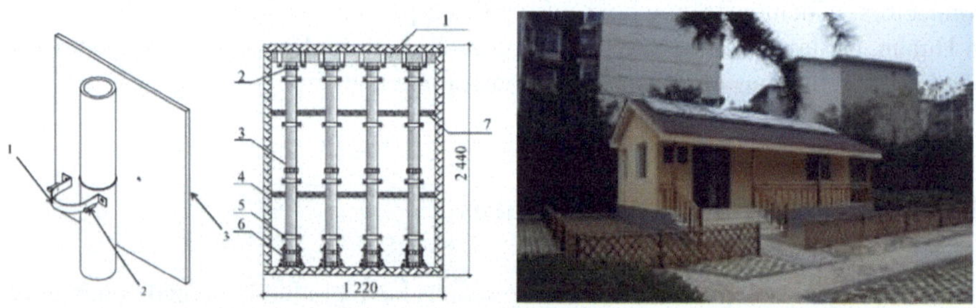

Fig. 6 Prefabricated round bamboo house with its shear wall

Xiaoxin used artistic rammed earth concrete as a new "soil" and bamboo scrimber as a new "wood" to translate the old traditional construction mode to express his respect to the ancients' wisdom in the construction of Zhaojun Museum (Fig. 7b). (1) In the renovation and expansion project of the Zen Center of Baoguo Temple (Fig. 7c), Architects Liang Jingyu and Ye Siyu tried to find light and small members

a. Youth Hostel in Niubeishan Mountain, Sichuan, China.

b. Zhaojun Museum, Huhhot, Inner Mongolia, China.

c. Zen Center of Baoguo Temple in Lezhi, Sichuan, China.

d. Outdoor Heteromorphic Bamboo Landscape of Shanghai Qinsen Group.

Fig.7 Application cases for structural uses of engineered bamboo. (Photos source: Fig. 7a provided by Hongyazhuyuan Science and Technology Company of China; Fig. 7b provided by China Construction Technology Group Co., Ltd and photo by Zhang Guangyuan; Fig. 7c provided by Approach Architecture Studio; Fig. 7d provided by Ganzhou Sentai Bamboo & Wood Co., Ltd.) [8]

to form a spatial grid structure of a large-span roof of the Zen Hall on the sites of the Sui Dynasty (581–618 A.D) buildings. Bamboo scrimber material was chosen and processed into 2 by 4 cm cross-section rods, solving the problem that this site can only rely on human transportation and small equipment. (2) The outdoor heteromorphic bamboo landscape of Shanghai Qinsen Group at the 2019 Beijing International Horticultural Exposition brought the application of glued laminated bamboo in the outdoor landscape into full play (Fig. 7d). The whole design shows the respect of the beauty of nature, with the inspiration of the curl and undulating three-dimensional space structure of wood shavings [8].

For those projects introduced as above, most applications belong to public buildings. Bamboo is also an excellent material for residential houses. A mature and reliable structure system is one of the critical issues for scaling it up. Lightweight GluBam frame structure is one of the most successful structural applications, and more than 20 lightweight GluBam buildings were built up to now (Fig. 8). Like lightweight wood-frame construction, the lightweight GluBam structure comprises the primary structural framing members and sheathing. It can be constructed following the platform-frame construction. The floor system is composed of joists

Fig. 8 Lightweight GluBam frame constructions

covered with sub-flooring to form a working surface upon which exterior walls and interior partition can be quickly erected. The shear walls are connected to the foundation or the lower story walls through the floor system by anchor connections. The roof system, normally prefabricated trusses, is connected to the shear walls' top beam by metal connections.

The lightweight GluBam frame structures have good lateral resistant behavior. The shear walls perpendicular to the lateral force can transfer the load to the horizontal diaphragms (floor or roof). The horizontal diaphragms distribute it to the shear walls parallel to the lateral force; finally, the shear walls transfer the force to the foundation. The lightweight GluBam shear wall is the main component to resist lateral loads. Figure 9 describes the fabrication process and its structural elements and connection system. Wall framing includes studs, top, and bottom beams. The vertical studs are supported on a bottom beam and support the top beam. Studs usually consist of 40 $\times$ 90 mm or 40 $\times$ 140 mm and are commonly spaced 305 or 405 mm on center. The stud spacing can be changed from 300 to 610 mm, depending on the load. The ply-bamboo panels are usually used as wall sheathing. The connection system consists of a panel-frame connection, frame-frame connection, and hold-down connection, which are the dominant factors determining the GluBam shear wall's lateral performance [9]. The high resistance of lightweight GluBam frame construction is provided when sheathing panels are adequately nailed to the framing members. Existing studies [10] indicate that lightweight GluBam shear walls have a high bearing capacity but limited ductile performance than wood-frame shear walls when using the same panel-frame connectors.

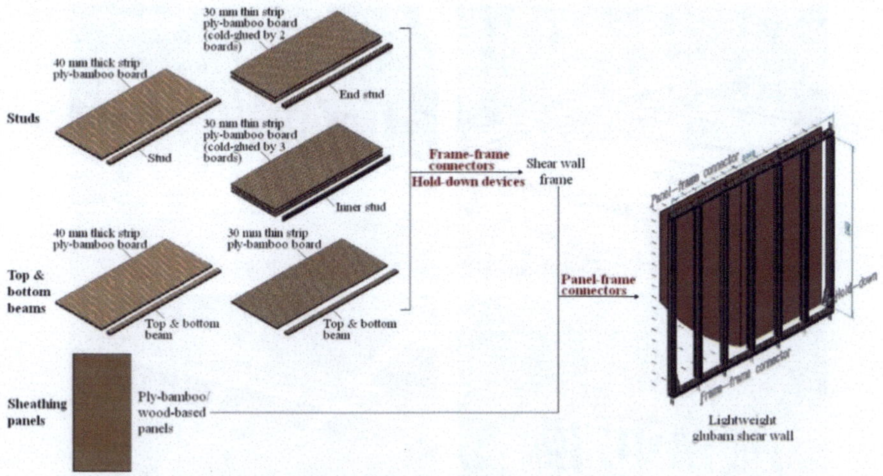

Fig. 9 The fabrication process of the GluBam shear wall

The lightweight GluBam structure floor consists of sills, girders, joists or floor trusses, and sub-flooring. The joists usually are spaced 305, 405, or 610 mm on centers and are supported by the foundation and the center girder. The size of the girders is determined by load calculation and detailing requirements. Sub-flooring or flooring sheathing is applied over the framing to provide a working platform, and the common sheathing material is a 15 mm thin strip ply-bamboo panel, as shown in Fig. 10. The roof system (Fig. 11) is typically made with prefabricated GluBam trusses as engineered trusses can reduce on-site labor and provide greater flexibility in interior walls' layout. Large-span GluBam trusses can be made up to 24 m long, as shown in Fig. 12.

4 Hybrid Bamboo Construction

High strength and modulus values of bamboo products on its main-fiber (x) direction support the application of bamboo materials, mainly as tension elements for hybrid bamboo structures such as bamboo-reinforced concrete structures. Due to the lack of steel for construction during World War I, hundreds of bamboo-reinforced concrete buildings were built in China, mainly in Canton. Some of them are still in service up to now as public buildings. However, due to biomaterials' durability limitations, applying bamboo reinforced concrete is mainly limited in research practices. Inspired by the recent development of cross-laminated timber (CLT), cross-laminated bamboo and timber (CLBT) was developed [11]. The CLBT is made by glue laminating under pressure the layers of laminated bamboo panels, and standard timber lumbers (SPF or fast-growing species in China). The orientation of the elements can be predetermined to satisfy design requirements. The CLBT construction is similar

Fig. 10 Processing and installation of lightweight GluBam shear wall and floor diaphragm

Fig. 11 Roof system in lightweight GluBam structure

Fig. 12 Large span GluBam truss

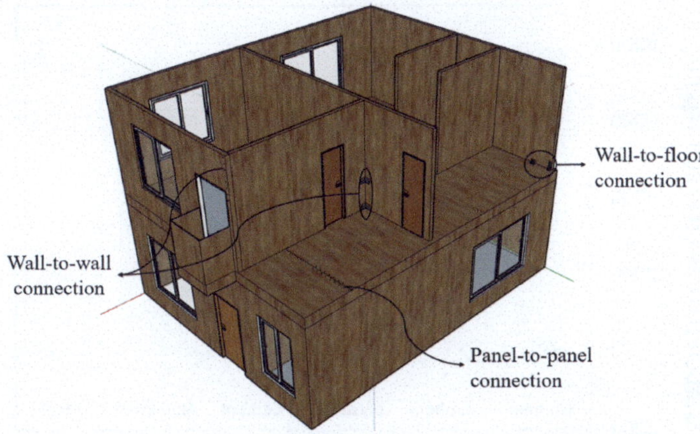

Fig. 13 Typical CLBT structures

to the lightweight frame structure; the difference is that the panel components are used instead of the framing members in lightweight GluBam structures, as shown in Fig. 13. The CLBT structure can be divided into prefabricated panel components; the horizontal floor (or roof) panels rest directly on top of vertical wall panels through simple metal connections. The CLBT structure system can be used in 3–4 story buildings or even high-rise structures. Due to the significant in-plane shear stiffness of the CLBT panel, the energy dissipation of structural components is provided by the anchorage connectors between the panel and the foundation (or floor) and the connection between adjacent panels. Therefore, the performance of these two types of connections determines the overall behavior of the CLBT structure.

Another successful application of engineered bamboo with other construction materials is the hybrid bamboo-steel space truss [12, 13], as shown in Fig. 14. The hybrid space truss system specially designed comprises two primary materials, glue-laminated bamboo and steel. The GluBam chords are mainly used for the upper compressive chords and web chords, while the steel pipes are mainly used for the bottom tensile chords. Compared with traditional steel space trusses, the hybrid space

Fig. 14 Hybrid bamboo-steel truss

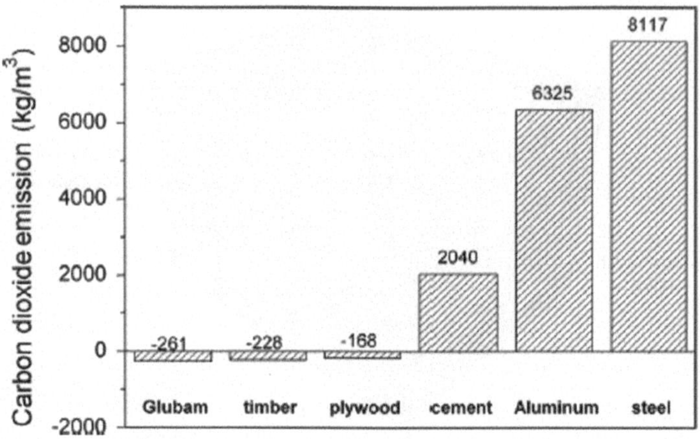

Fig. 15 Carbon dioxide emission of GluBam and comparable construction materials [3]

truss system is developed with the concept of achieving possible maximum carrying capacity and lighter self-weight together.

5 Carbon Emission of Bamboo Construction

As a structural material, its influence on the environment is crucial in today's trend towards a sustainable construction industry. The low carbon emission of bamboo products has been widely recognized, making it very attractive in today's move towards a sustainable society. Numerous studies have shown that bamboo forests are more superior to other forests in capturing CO_2. The histogram shown in Fig. 15 compares Glubam with other construction materials, including timber, plywood, cement, aluminum, and steel, in carbon emission [3]. Cement, aluminum, and steel discharge much more carbon dioxide than the other materials. In comparison, GluBam is carbon emission negative and outperforms timber and plywood.

6 Conclusion

The development in scientific research and engineering practices on bamboo materials and constructions has been noticed in the recent 30 years. The promising potentials of bamboo materials used in the construction industry can be observed clearly based on the accumulated research goals and construction examples. The structural bamboo can be modeled as a unidirectional reinforced bio-composite material. High tension and compression strength values can be noticed for it. However, relatively low modulus values, around 10 GPa, can be noticed for bamboo. Thus, the design of bamboo structures is typically governed by deformation limitations. The hybrid timber/steel-bamboo structure is an effective solution for optimizing bamboo material usages in construction. The carbon emission of bamboo construction, using GluBam structures as an example, is discussed and compared with timber, plywood, cement, aluminum, and steel structures. Superior carbon emission performances can be noticed for bamboo construction. More research on standardization and industrial-grade methods is suggested for the vast application of this environmentally friendly material.

References

1. Xiao, Y., Z. Li, and K.W. Liu. 2020. *Modern engineered bamboo structures*. The Netherlands: CRC Press/Balkema.
2. Wang, J.S., C. Demartino, Y. Xiao, and Y.Y. Li. 2018. Thermal insulation performance of bamboo- and wood-based shear walls in light-frame buildings. *Energy and Buildings* 168: 167–179. https://doi.org/10.1016/j.enbuild.2018.03.017.
3. Xiao, Y., R.Z. Yang, and B. Shan. 2013. Production, environmental impact and mechanical properties of glubam. *Construction and Building Materials* 44: 765–773. https://doi.org/10.1016/j.conbuildmat.2013.03.087.
4. Yang, R.Z., Y. Xiao, and F. Lam. 2014. Failure analysis of typical glubam with bidirectional fibers by off-axis tension tests. *Construction and Building Materials* 58: 9–15. https://doi.org/10.1016/j.conbuildmat.2014.02.014.
5. Xiao, Y., Y. Wu, J. Li, and R.Z. Yang. 2017. An experimental study on shear strength of glubam. *Construction and Building Materials* 150: 490–500. https://doi.org/10.1016/j.conbuildmat.2017.06.005.
6. Li, Z., J. Zhang, R. Wang, G. Monti, and Y. Xiao. 2020. Design embedment strength of plybamboo panels used for glubam. Journal of Materials in Civil Engineering 32 (5). https://doi.org/10.1061/(asce)mt.1943-5533.0003128.
7. Li, Z., X.Z. He, Z.M. Cai, R. Wang, Y. Xiao. 2021. Mechanical properties of engineered bamboo boards for glubam structures. Journal of Materials in Civil Engineering 33 (5). https://doi.org/10.1061/(asce)mt.1943-5533.0003657.
8. Liu, K.W., Q.F. Xu, G. Wang, F.M. Chen, and Y.B. Leng. 2019. *Contemporary bamboo architecture in China*. Beijing: China Architecture & Building Press.
9. Wang, R., S.Q. Wei, Z. Li, and Y. Xiao. 2019. Performance of connection system used in lightweight glubam shear wall. *Construction and Building Materials* 206: 419–431. https://doi.org/10.1016/j.conbuildmat.2019.02.081.
10. Wang, R., Y. Xiao, and Z. Li. 2017. Lateral loading performance of lightweight glubam shear walls. Journal of Structural Engineering 143 (6). https://doi.org/10.1061/(asce)st.1943-541x.0001751.
11. Xiao, Y., H. Cai, and S.Y. Dong. 2021. A pilot study on cross-laminated bamboo and timber beams. *Journal of Structural Engineering* 147 (4): 06021002.
12. Wu, Y., and Y. Xiao. 2018. Steel and glubam hybrid space truss. *Engineering Structures* 171: 140–153. https://doi.org/10.1016/j.engstruct.2018.05.086.
13. Quaranta, G., C. Demartino, and Y. Xiao. 2019. Experimental dynamic characterization of a new composite glubam-steel truss structure. Journal of Building Engineering 25. https://doi.org/10.1016/j.jobe.2019.100773

A Review on Round Bamboo Structural Applications and Perspectives

Alessandra Simi, Roya Mostaed Shirehjini, Da Shi, Cristoforo Demartino, and Zhi Li

Abstract Nowadays, bamboo as a readily available, inexpensive, renewable raw material is increasingly used in construction in several countries around the world. In this context, the use of bamboo culm for structural applications is one of the key concepts increasingly grown over the past few decades. This is motivated by the fact that regardless of being a rough-and-ready (low-tech) material, round bamboo culms are found to be easy to implement and capable of yielding architectural and structurally high-performing light-weight constructions and building designs. In this study, we aim to review design approaches for round bamboo structures, to provide a recent state-of-the-art of the relevant scientific and technical literature.

Keywords Bamboo culm · Structural bamboo · Connections · Structural forms

1 Introduction

In the past few decades, bamboo-based building materials have progressively suggested new promising approaches in sustainable construction. Bamboo is inherently renewable, with a great potential to provide carbon–neutral raw materials in an inexhaustible system, thus providing significant environmental and economic advantages. Moreover, it is considered as the fastest growing plant species, with a growth rate that can reach up to 100 cm in a day.

As a building material, bamboo culm has been shown to have remarkable mechanical properties. In addition, it is characterized by a short rotation age: Hisham [1] concluded that a 3.5-year-old bamboo is a suitable material for construction.

A. Simi · R. M. Shirehjini
College of Civil Engineering, Nanjing University of Technology, Nanjing, Jiangsu, China

D. Shi · C. Demartino (✉)
Zhejiang University-University of Illinois at Urbana Champaign Institute, Zhejiang University, Haining, Zhejiang, China
e-mail: cristoforodemartino@intl.zju.edu.cn

Z. Li
College of Civil Engineering and Architecture, Zhejiang University, Hangzhou, Zhejiang, China

K. Liu et al. (eds.), *2019 International Bamboo Construction Competition*, Springer Tracts in Civil Engineering, https://doi.org/10.1007/978-3-030-91990-0_3

Bamboo culm has a cylindrical, hollow, anisotropic body, which makes it particularly challenging to deal with [2]. During less than three decades, there has been a significant growth in research and development on bamboo culm structures. In this review we aim to review bamboo culm application in construction and the relevant findings of previous works.

1.1 Constructing with Bamboo

Today, bamboo is a primary housing material for over one billion people [3]. Bamboo is largely used because it is cheap, locally available, has high disaster-tolerance ability due to being strong, lightweight, flexible, environmentally compatible. Stilt bamboo houses scattered in ASEAN countries and bamboo houses in South American countries are good examples of such phenomena [4]. Bamboo culms endure more lateral forces, this can mitigate damages during floods, typhoons, earthquakes. Buildings made of bamboo have been left intact whereby many concrete and masonry buildings were severely damaged during the devastating earthquakes that took place in Costa Rica (1991, Mw 7.6), Colombia (1999, Mw 6.4), Ecuador (2016, Mw 7.8) and Indonesia (2018, Mw 6.9); this can be encouraging for bamboo applications in earthquake-prone areas.

Nevertheless, despite such modern progress in this area and the successful results achieved, it must be said that, as of today, bamboo construction has been more related to those environmentally concerned communities. Indeed, bamboo culm as a structural or building material is still relatively unknown to many decision makers and clients. Besides, there is little standardization and few design criteria, which could challenge the designers and lead to loss of confidence and acceptance among them.

1.2 Mechanical Properties

No two bamboo culms are completely alike; indeed, both physical and mechanical characteristics of bamboo culms, can be unique for each bamboo species and even for every single culm. Thus, the mechanical properties should be related to the physical characteristics of bamboo culms.

Table 1 shows approximate values of key mechanical properties of some structural bamboo species observed and recorded by different authors.

Table 1 Literature mechanical properties of some structural bamboo species

Bamboo species	TS (MPa)	CS (MPa)	SS (MPa)	BS (MPa)	References
Guadua angustifolia	143	40–43	8	209	[5]
Phyllostachys pubescens	115–309	48–60	11–15	9–27	[6, 7]
Phyllostachys edulis	129–193	69	15	–	[8–10]
Dendrocalamus asper	232	73	7	20	[11]
Bambusa vulgaris	232	78	8.5	20	[11]

TS = Tensile Strength; CS = Compressive Strength; SS = Shear Strength; BS = Bending Strength

2 Design and Implementation

2.1 Modeling and Form-Finding

Dealing with a round bamboo culm is affected by uncertainty and ambiguity. The difficulty of constructing with bamboo often lies in the wide differences in structural and dimensional features among different bamboo culms and even within the same culm. This makes bamboo culms dependent on many variables, which require time-consuming measurements and complex computational procedures. Consequently, this may lead to a lack of confidence in designing suitable forms and geometries.

To carry out physical modelling in structural design, many studies have suggested to ease geometrical and structural analysis and to enhance computational accuracy [12]. Once the detailed structural models and their form are well-understood, they are transformed into a digital model. The approach is twofold: either physical modelling or digital simulation.

Physical modelling has been used to identify optimum assembly plans and to generate functional forms of structures [13, 14]. The physical form-finding method has been also used extensively in active-bending bamboo structures (e.g., domes, arches, tensile grid shells) [15, 16]. This approach is only based on experimental form-finding without numerical calculations.

As for the digital simulation approach, two powerful and commonly used digital tools are Grasshopper® and Rhinoceros® with Kangaroo and Karamba3D [17] plugin for computational simulation and form-finding of bamboo structures [18–20]. These programs resemble finite element analysis methods [5, 16, 21, 22].

2.2 Joints

The approach used to construct and implement bamboo structures is comparable with that of timber; joining lightweight structural components together by means of connectors. Joints (or connectors) as critically important aspect of bamboo structures are to ensure an efficient load bearing system or otherwise it can be the main cause of failure in structural stability and performance [23].

Lashing or ropework can be considered as most widely used bamboo connecting method. Despite being an old technique, lashing is being used in assembling several modern bamboo structures. The ropes used for lashing purpose can be made of natural fibers such as hemp, lime bast, coir, rattan and bamboo, or synthetic fibers such as polyester cords.

Bolts, as simple but strong and efficient mechanical fastening, offer opportunities that traditional lashing joints do not. This technique requires drilling holes on the wall of the bamboo culms and to use bolts and nuts to fasten and secure the joints between the bamboo components. Different forms of bolted bamboo connections have been proposed: using multi-bolted joint [23–25], or other supporting means such as hooks [23], mounting inner U-shaped metal plates [26], natural fiber straps [27] to secure the bolts from loosening and disengaging.

Clamped joints are an interesting alternative connection system to bolted ones. Adding clamps (made of steel or plywood) to the outer circumference of the culm wall increases the load capacity of the joints through friction contact and improving force distribution [28].

Hubs are node-like connecting members with radially oriented elements at angles, used for geodesic dome construction. The hubs can be from different materials (e.g., steel, PVC) or connection technologies. Some hubs have pipe connector ends embedded inside each culm and then bolted [29, 30]; other hubs are with steel brackets or rods inserted into the bamboo culms and firmly anchored (Richard et al. 2017), [30].

2.3 Foundations and Plinths

In the last two decades, many innovative ways have been developed to enhance the structure stability and firmness at the foundation level, using concrete and steel [31]. In Ref. [32], the bamboo columns are embedded in sand-packed bags or buckets, which allow them to progressively sink into the sand, thus enhancing the foundational constraint. Such approach is suitable for lightweight structures, such as pavilions. Precast concrete plinths (usually in frustum form, see Fig. 1) are laid on the ground to support each column unit [33]. In heavy bamboo structures, resting the plinth and foundation upon hard surfaces such as bedrock, concrete or compacted soil is required along with laying a concrete slab or strip foundation (of concrete or brick) [34].

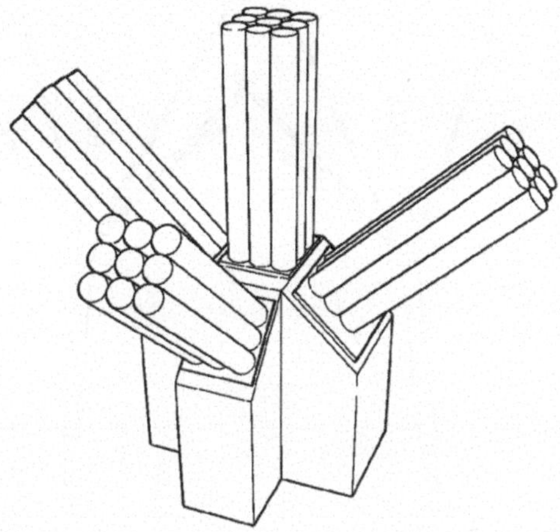

Fig. 1 Bundled bamboo culms branching out from a raised concrete plinth

3 Structural Systems and Forms

3.1 Structural Components: Columns, Bearing Walls, Floors and Roofs

Bamboo culms can act as structural column when stood up individually on a short center-to-center distances [34, 35]. However, in modern bamboo structure, multiple bamboo culms in bundled and V-shaped configurations [32] have been used to form bamboo culms, with longer spans and loads. Bundled bamboo columns can be configured in a vertically upright position [35–37], curved [38], splayed out at various angles and form [39–42], in fusiform shape, twisted cylinder, or a combination of them [32] (Figs. 2 and 3).

As for bearing walls, one traditional but effective lightweight load-bearing bamboo wall is bahareque system; it is a composite bamboo-framed wall with bamboo strips, chicken wire, adhesive materials such as soil–cement or mortar cement used for filling and plastering the wall [43].

Bamboo has also been used in flooring applications of upper levels in modern bahareque system housing [43]. In this system, bamboo culms with large diameters are used as secondary beams (joists), woven bamboo mats coated with cement mortar for floor finishes.

Round bamboo has been also successfully used as a main structural component in many advanced pedestrian bridges. Using pre-tensioned bamboo arches for setting camber is the key for constructing bamboo bridges [44].

Thanks to the advantages offered by its high strength-to-weight ratio, bamboo has been considered as potentially valuable material for load-bearing roof systems (e.g., truss, arched and dome) when operatively interconnected [45]. Vengala et al. [33] have demonstrated a good example of bamboo truss with connections of bamboo plates to reduce shearing and splitting risk. In an exhaustive study on curved bamboo

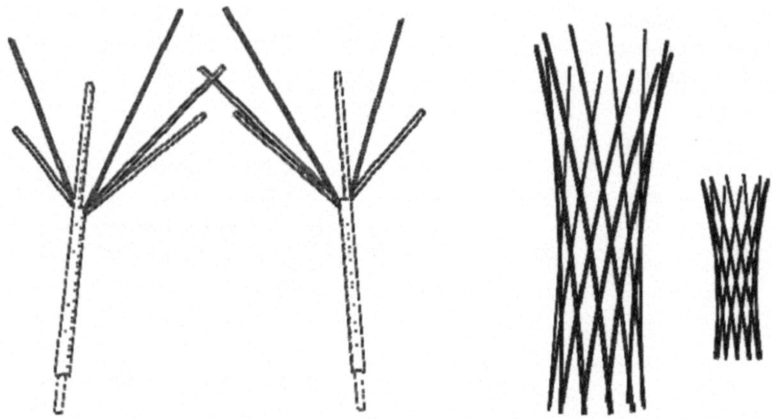

Fig. 2 Two different bamboo column configurations: (left) tree-like bamboo columns and (right) bamboo column with spiral base

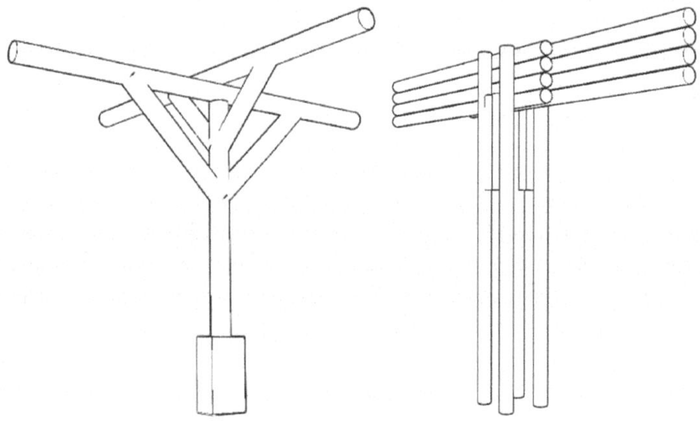

Fig. 3 Bamboo columns and beams: (left) tree-like knee-bracing, (right) configuration of bamboo columns and beams [37]

members by Tahmasebinia et al. [22], arched bamboo structures such as barrel structures with cross-over design used less bamboo culms and provided greater load carrying capacity than the straight arrangement of the arches.

3.2 Mainstream Structural Forms: Form-Active Structures

Form-active structures are subject to pure tensile (e.g. cables and membranes) or compressive (arches) external loading, without shear or bending moments. Therefore, bamboo culm could hold particular promise for this structural families due

Fig. 4 Cable-stayed bamboo cantilever, toll gate in Armenia, Colombia. Photo by: Forgemind ArchiMedia [48]

to its considerable tensile and compressive strength. In this context, Stamm [46], used bamboo culms effectively in construction of the Bali Bridge, composed of two bamboo arches linked together by steel bolts and bamboo-made dowels.

Bamboo-cable composite by Fu et al. [47] is proposed and demonstrated as a favorable application of bamboo in cable structures. The Páez Bridges, the Cucuta Bridge are another successful cable structures developed by Stamm [46], in which bamboo culms play as tensile member. Moreover, bamboo culms have been used effectively in cable-stayed cantilevers of the toll gate structure in Armenia, Colombia [48] (Fig. 4).

With the developments in connections, a number of aesthetic bamboo residential buildings have been constructed worldwide [37, 49]. Energy Efficient Bamboo House [50] is a two-storey bamboo framed house on stilts with "industrialized construction system", using novel aluminum connection system whereby structural bamboo components can be detached and replaced (Fig. 5).

4 Concluding Remarks

Raw bamboo culm is receiving new appreciation as an alternative material for construction today; according to the literature, unprocessed (or minimally processed) bamboo culm has advantages of high strength-to-weight ratio and flexibility, while being characterized by the least level of energy and pollution intensity. In view of this, many researchers and construction practitioners have been making great efforts to identify and develop reliable and effective approaches and strategies to construct with bamboo culms, in recent three decades.

Following this, a spectacular progress has been made in bamboo construction during this short time, bamboo culm material has shown an extensive scope of structural applications in modern construction. Bamboo culms have been adopted effectively as tensile and compressive load supports in different structural systems such as

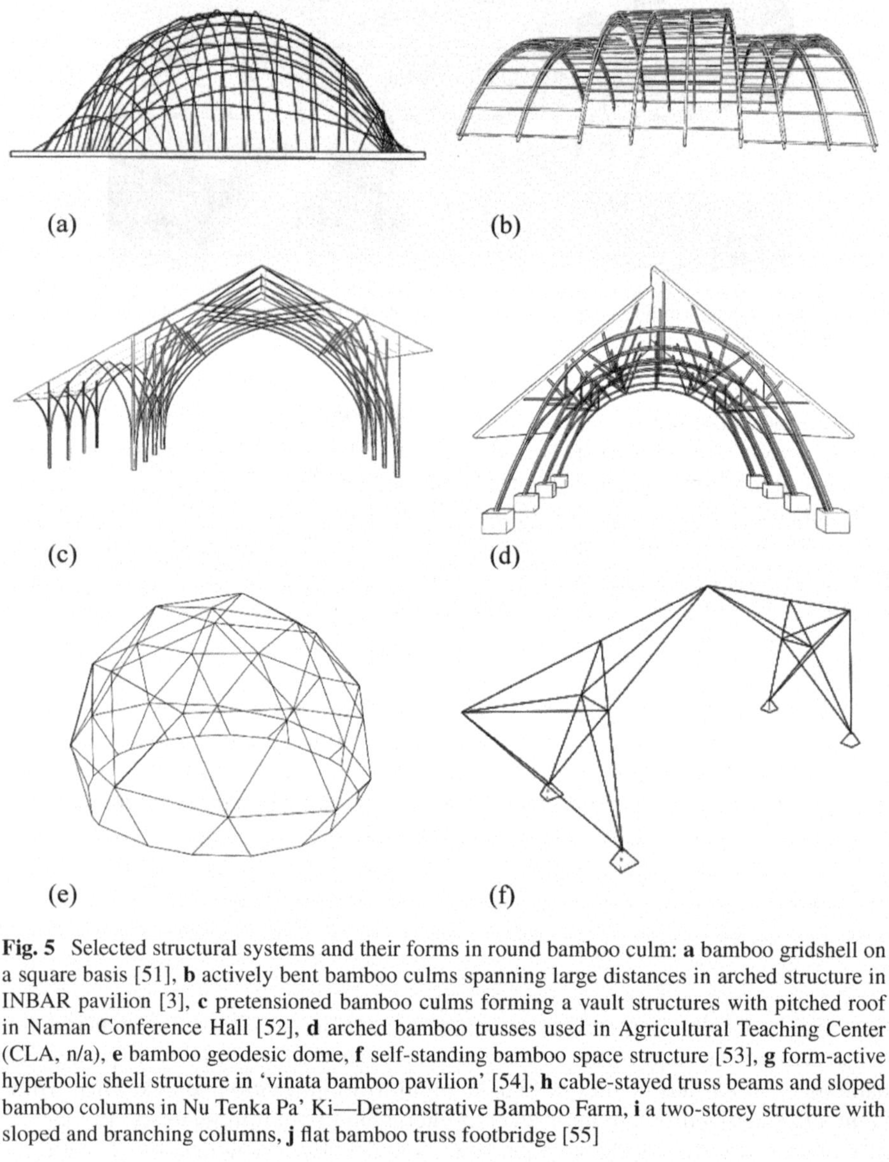

Fig. 5 Selected structural systems and their forms in round bamboo culm: **a** bamboo gridshell on a square basis [51], **b** actively bent bamboo culms spanning large distances in arched structure in INBAR pavilion [3], **c** pretensioned bamboo culms forming a vault structures with pitched roof in Naman Conference Hall [52], **d** arched bamboo trusses used in Agricultural Teaching Center (CLA, n/a), **e** bamboo geodesic dome, **f** self-standing bamboo space structure [53], **g** form-active hyperbolic shell structure in 'vinata bamboo pavilion' [54], **h** cable-stayed truss beams and sloped bamboo columns in Nu Tenka Pa' Ki—Demonstrative Bamboo Farm, **i** a two-storey structure with sloped and branching columns, **j** flat bamboo truss footbridge [55]

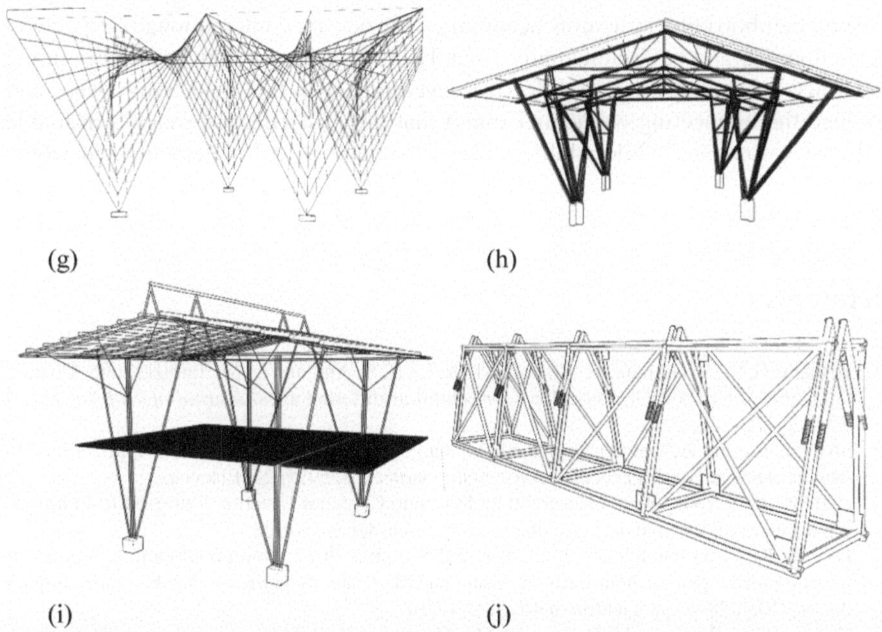

Fig. 5 (continued)

trusses, space and frame structures. In addition to this, having high modulus of elasticity and a tubular structure, bamboo material can smoothly adapt to be prestressed and employed in active bending structures such as arches and shells.

A significant part of the progress made in bamboo construction is due to the advanced methods and technologies adopted in bamboo construction in the evolution of the concept of constructing with bamboo.

Introducing digital form-finding and analysis into the design procedure with an emphasis on physical modelling, has enabled the designers to achieve functional structures and highly sophisticated forms, as well as controlling their realization.

Also noticeable is the advancement made in bamboo joints to improve the behavior of bamboo structures, their stability, forms. Depending on the bamboo structural system and loads, either traditional or innovative joint technologies, or blending of both may be employed. However, the existing joints have been criticized for not being capable of meeting the needs for the high load bearing capacity of the connections with economical operation; on the other hand, more research is to be undertaken to amend adverse characteristics of bamboo culm such as brittleness with appropriate joint design [23].

With bamboo culm structures becoming more accepted internationally especially as economical and environmentally friendly alternatives for conventional building materials, a key step towards effective development in this field is to design and produce the connecting elements (joints) that are cost-effective, readily available and easy-to-operate, while adding stiffness to bamboo culm elements and whole structure.

References

1. Hisham, H.N., S. Othman, H. Rokiah, M. A. Latif, S. Ani, and M.M. Tamizi. 2006. Characterization of bamboo Gigantochloa scortechinii at different ages. *Journal of Tropical Forest Science* 236–242.
2. Sharma, B., and A. van der Vegte. 2020. Engineered bamboo for structural applications. In *Nonconventional and vernacular construction materials*, 597–623. Elsevier.
3. INBAR. 2019. INBAR pavilion architect Mauricio Cardenas Laverde. Retrieved from https://www.inbar.int/inbar-pavilion-architect-mauricio-cardenas-laverde.
4. Das, P., K. Chaaruchandra, P. Sudhakar, and S. Satya. 2012. Traditional bamboo houses of north-eastern region: A field study of Assam and Mizoram. *Key Engineering Materials*. https://doi.org/10.4028/www.scientific.net/KEM.517.197.
5. García, B., C. Preciado, M. Bedoya, and O. Mendoza. 2019. Mechanical characterization of *h* and Guadua bamboo. In *MATEC Web of Conferences*, eds. C. Chastre, and P. Mendonça, 303, 03002. https://doi.org/10.1051/matecconf/201930303002.
6. Deng, J., F. Chen, G. Wang, and W. Zhang. 2016. Variation of parallel-to-grain compression and shearing properties in Moso bamboo culm (Phyllostachys pubescens). *BioResources* 11: 1784–1795.
7. Yu, H.Q., Z.H. Jiang, C.Y. Hse, and T.F. Shupe. 2008. Selected physical and mechanical properties of Moso Bamboo (Phyllostachys pubescens). *Journal of Tropical Forest Science* 6.
8. Schneider, N., W. Pang, M. Gu. 2014. Application of bamboo for flexural and shear reinforcement in concrete beams. In Structures Congress 2014, 1025–1035. Boston, Massachusetts, United States: American Society of Civil Engineers. https://doi.org/10.1061/9780784413357.091.
9. Molari, L., L. Mentrasti, and M. Fabiani. 2020. Mechanical characterization of five species of Italian bamboo. *Structures* 24: 59–72. https://doi.org/10.1016/j.istruc.2019.12.022.
10. Akinbade, Y., K.A. Harries, C.V. Flower, I. Nettleship, C. Papadopoulos, and S. Platt. 2019. Through-culm wall mechanical behaviour of bamboo. *Construction and Building Materials* 216: 485–495.
11. Awalluddin, D., M.A. Mohd Ariffin, M.H. Osman, M.W. Hussin, M.A. Ismail, H.-S. Lee, and N.H. Abdul Shukor Lim. (2017). Mechanical properties of different bamboo species. In *MATEC Web of Conferences*, eds. J.-W. Park, H. Ay Lie, H. Hardjasaputra, P. Thayaalan, 138, 01024. https://doi.org/10.1051/matecconf/201713801024.
12. Kamath, A.V. 2013. Digitally designed architectural form-built using craft-based fabrication–weaving a complex surface as a bamboo reticulated shell. In Proceedings of the 18th international conference on computer-aided architectural design research in Asia CAADRIA, 623–632. Singapore.
13. Seixas, M., J. Bina, P. Stoffel, J.L. Ripper, L.E. Moreira, and K. Ghavami. 2017. Active bending and tensile pantographic bamboo hybrid amphitheater structure. *International Association for Shell and Spatial Structures (IASS)* 58: 239–252. https://doi.org/10.20898/j.iass.2017.193.872.
14. Crolla, K. 2018. Bending bamboo rules: Beyond century-old typologies. *Journal of Architectural Education* 72: 135–145.

15. Seixas, M.A., J.L. Ripper, and K. Ghavami. 2015. Construction of mobile bamboo dome covered with textile canvas. In *16th Non conventional building material and technologies international conference*, Winnipeg, Canadá.
16. Seixas, M., L.E. Moreira, J. Bina, and J.L. Ripper. 2018. Design and analysis of a self-supporting bamboo roof structure applying flexible connections. In *Proceedings of IASS annual symposia 2018*, 1–9. International Association for Shell and Spatial Structures (IASS).
17. Michiels, T., L. Lu, R. Archer, S. Adriaenssens, and G. Tresserra. 2017. Design of three hypar roofs made of Guadua bamboo. *Journal of the International Association for Shell and Spatial Structures* 58: 95–104.
18. Panagoulia, E., and S. Schleicher. 2016. Bending-active structures a case study for an Office Chaise Lounge.
19. Crolla, K., and A. Fingrut. 2016. Protocol of error: The design and construction of a bending-active gridshell from natural bamboo.
20. Hansen, L.H., and S. Kim. 2016. 'COCOON': A bamboo building with integration of digital design and low-tech construction. In *International conference on structures and architecture.*
21. Ho, G.W.-M., and M. Fong. 2018. Design and construction of curved bamboo structure. In *Ninth international conference on advances in steel structures*. Hong Kong, China: The Hong Kong Institute of Steel Construction.
22. Tahmasebinia, F., R. McDouga, S. Sepasgozar, E. Abberton, G.H. Joung, M.P. Joya, S.M.E. Sepasgozar, and F.A. Marroquin. 2020. Development of preliminary curved bamboo member design guidelines through finite element analysis. *Sustainability* 12: 822. https://doi.org/10.3390/su12030822.
23. Hong, C., H. Li, R. Lorenzo, G. Wu, I. Corbi, O. Corbi, Z. Xiong, D. Yang, and H. Zhang. 2019. Review on Connections for Original Bamboo Structures. *Journal of Renewable Materials* 7: 713–730. https://doi.org/10.32604/jrm.2019.07647.
24. Sonar, I.P., and V.R. Siddhaye. 2009. Theoretical and experimental investigation on single and double bolted bamboo joint under axial tension. *Journal of Structural Engineering* 36.
25. Hu, H., J. Yang, F.L. Wang, and Y.M. Zhang. 2018. Mechanical properties of bolted joints in prefabricated round bamboo structures. *Journal of Forestry Engineering* 3: 128–135.
26. Tan, G.Y., and L. Yang. 2014. *Tectonics of bamboo.* Nanjing: Southeast University Press.
27. Moreira, L.E., and K. Ghavami. 2012. Limits states analysis for bamboo pin connections. *Key Engineering Materials* 517, 3–12.
28. Masdar, A., B. Suhendro, S. Siswosukarto, and D. Sulistyo. 2015. The study of wooden clamps for strengthening of connection on bamboo truss structure. *Jurnal Teknologi* 72
29. Vahanvati, M. (2015). The challenge of connecting bamboo. In *10th World Bamboo Congress*, Damyang.
30. Wang, T.-H., O. Espinosa Trujillo, W.-S. Chang, and B. Deng. 2017. Encoding bamboo's nature for freeform structure design. *International Journal of Architectural Computing* 15: 169–182. https://doi.org/10.1177/1478077117714943.
31. Correal, J.F. 2016. Bamboo design and construction. In *Nonconventional and vernacular construction materials*, eds. K.A. Harries, B. Sharma, 393–431. Woodhead Publishing. https://doi.org/10.1016/B978-0-08-100038-0.00014-7.
32. Minke, G. 2012. *Building with bamboo: Design and technology of a sustainable architecture.* Walter de Gruyter.
33. Vengala, J., H.N. Jagadeesh, and C.N. Pandey. 2008. Development of bamboo structure in India. In *Modern bamboo structures*, 63–76. CRC Press. https://doi.org/10.1201/9780203888926-11.
34. Jayanetti, D.L., and P.R. Follett. 2008. Bamboo in construction. In *Modern bamboo structures*, 35–44. CRC Press.
35. Paudel, S. K. (2008). Engineered bamboo as a building material. In *Modern bamboo structures*, 45–52. CRC Press.
36. House, E.E. 2017. Energy Efficient Bamboo House/Studio Cardenas Conscious Design. Retrieved from http://www.archdaily.com/868926/energy-efficient-bamboo-house-studio-cardenas-conscious-design.

37. Architects, V. 2014. Son La Restaurant/VTN Architects. Retrieved 08 Apr 2020, from ArchDaily: https://www.archdaily.com/559125/son-la-restaurant-vo-trong-nghia-architects.
38. Nurdiah, E.A. 2016. The potential of bamboo as building material in organic shaped buildings. *Procedia - Social and Behavioral Sciences* 216: 30–38. https://doi.org/10.1016/j.sbspro.2015.12.004.
39. Supreedee, R. 2015. *Research, education, design in Thai bamboo architecture*. Damyang: Curran Associates Inc.
40. Markham, D. 2019. *Meet the woman building stunning sustainable homes from bamboo*. Retrieved from https://www.treehugger.com/green-architecture/meet-woman-building-stunning-sustainable-homes-bamboo.html.
41. Schröder, S. 2019. *The Reality about building with bamboo*. Retrieved from https://www.guaduabamboo.com/construction/the-reality-about-building-with-bamboo.
42. Hudson, D. 2012. *PT bamboo pure: Green school*, Bali. Retrieved from https://www.designboom.com/architecture/pt-bamboo-pure-green-school-bali/.
43. Kaminski, S., A. Lawrence, and D. Trujillo. 2016 Design guide for engineered bahareque housing. In *INBAR–International network for bamboo and rattan*, Beijing, China.
44. Carmiol-Umaña, V. 2016. Guadua bamboo in pedestrian bridges. *Revista Tecnología en Marcha* 29: 3–13.
45. Syeda, A., B.J. Shrujal, and J. Kumar. 2014. A case study on bamboo as green building material. *International Journal of Engineering and Advanced Technology* 4: 78–82.
46. Stamm, J. 2009. Seven concepts to build a bamboo bridge. *Proceeings of VIII World Bamboo Congress ISSN* 2150: 70–100.
47. Fu, Y., L. Hu, Y. Hu, and X. He. (2014). Study on the innovative application of bamboo-cable composite structures. UIA2014 Durban architecture otherwhere.
48. Sävfors, I. 2020. Echo-building. Retrieved 8 Apr 2020, from Saevfors Consulting: http://www.saevfors.se/Eco_Building.html.
49. Griffiths. (2014). *Low-cost Ecuador house built with rainforest wood and bamboo*. Retrieved from https://www.dezeen.com/2014/12/14/convento-house-enrique-mora-alvarado-bamboo-low-cost-ecuador/.
50. Laverde, M.C. 2016. *Energy efficient bamboo house*. Retrieved from https://www.studiocardenas.it/index.php/en/rainwater/32-thermal-control/148-energy-efficient-bamboo-house.
51. Correia de Melo, J.V., J.L. Ripper, and L.E. Moreira. 2013. Development of bamboo structures based on minimal surfaces and natural formations. In *Proceedings of the IASS annual symposium: Beyond the limit of man*.
52. VTN Architects. 2016. *Naman Conference Hall*. Retrieved from VTN Architects.
53. Moreira, L.E., M. Seixas, J. Bina, and J.L. Ripper. 2017. Self-supporting bamboo space structure with flexible joints. In *Proceedings of the 17th IC NOCMAT*, Mexico.
54. Barandy, K. (2019). *VTN architects introduces the hyperbolic vinata bamboo pavilion to urban Hanoi*. Retrieved from Deisgnboom: https://www.designboom.com/architecture/vtn-architects-vinata-bamboo-pavilion-01-17-19/.
55. Paraskeva, T., N.P.N. Pradhan, C.D. Stoura, and E.G. Dimitrakopoulos. 2019. Monotonic loading testing and characterization of new multi-full-culm bamboo to steel connections. *Construction and Building Materials*, 210, 473–483.
56. Richard, M.J., Gottron, J., Harries, K.A., & Ghavami, K. 2017. Experimental evaluation of longitudinal splitting of bamboo flexural components. *Proceedings of the Institution of Civil Engineers-Structures and Buildings* 170(4): 265–274.

INBAR Garden Pavilion: Bamboo Eye

Mauricio Cárdenas Laverde

Abstract Since construction is one of the biggest culprits to global greenhouse emissions, introduce bamboo into the building market chain is urgent at this particular time in history. INBAR Garden Pavilion named Bamboo Eye, located at the 2019 Beijing International Horticultural Exposition, is a demonstration of an innovative contemporary architectural solution for real problems of the environmental crisis from perspectives of design, construction, and life cycle. Bamboo Eye sets as an example of the potentiality of using round bamboo for construction as it demonstrates its structural properties, pushes forward the studies of international bamboo construction standards and inspires a new contemporary architectural language. Bamboo Eye is the largest round bamboo structure in Northern China.

Keywords Round bamboo structure · Bamboo arch structure · Bamboo garden · Bamboo weaving

1 Concept Design

INBAR Garden Pavilion known as Bamboo Eye, located at the 2019 Beijing International Horticultural Exposition, is a contemporary interpretation of the traditional concept of "a Pavilion in the Garden" by fusing the pavilion into the Garden, creating one single element: a Garden-Pavilion, an innovative sustainable typology where architecture, engineering and landscape melt into each other. Garden and Pavilion are one and connected through the fluidity of naturally moving from the exterior towards the interior space. The work of the contemporary artist Lucio Fontana, the cuts on the canvas, inspired the conceptual exploration and the research for the architectural form (Fig. 1). The strong symbiosis between architecture and landscape is generated through soft shapes and forms.

Since the very beginning, parallel to the study of the form and functional layout, passive design strategies were taken into consideration (Fig. 2). We considered the

M. C. Laverde (✉)
Studio Cardenas Conscious Design, Milan, Italy
e-mail: m.cardenas@studiocardenas.it

K. Liu et al. (eds.), *2019 International Bamboo Construction Competition*, Springer Tracts in Civil Engineering, https://doi.org/10.1007/978-3-030-91990-0_4

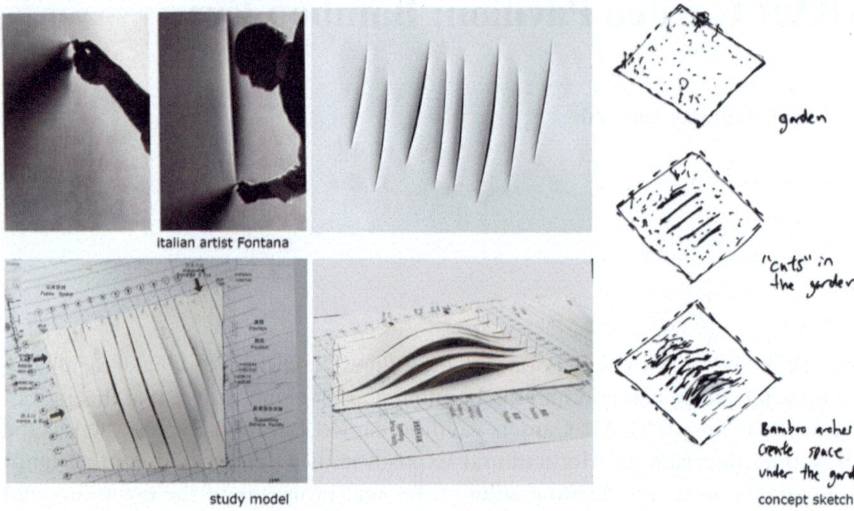

Fig. 1 Artist Lucio Fontana is working; sketches by Mauricio Cardenas Laverde; Study models by Studio Cardenas Conscious Design

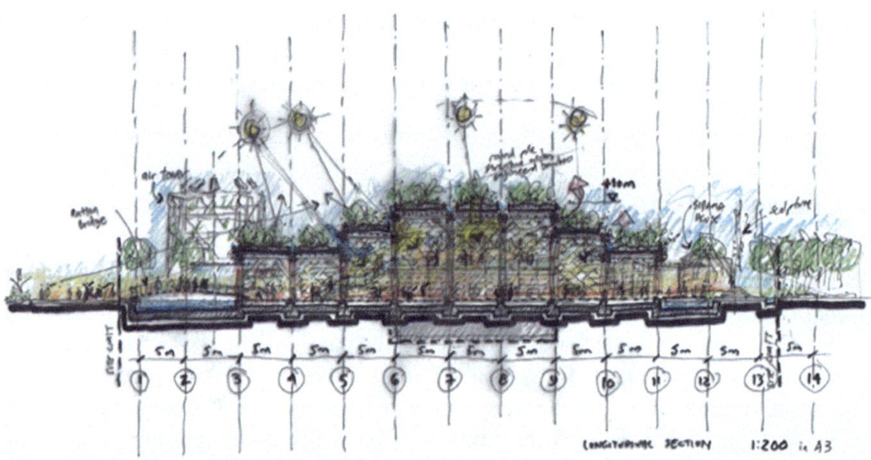

Fig. 2 Sketch by Mauricio Cardenas Laverde, showing first ideas of passive design

use of tangible and intangible natural materials such as natural light, breeze, water, pebbles, soil, and plants. The interior space under the garden is filled with natural light and refreshed through natural ventilation. Natural light, as the breeze, are essential intangible materials part of the design [1].

The bearing structure, which detaches the garden soil from the ground, is made of natural bamboo poles, a perfect product of nature that best integrates the values of ecology, community, nutrition, and beauty [2] (Fig. 3). The series of natural bamboo

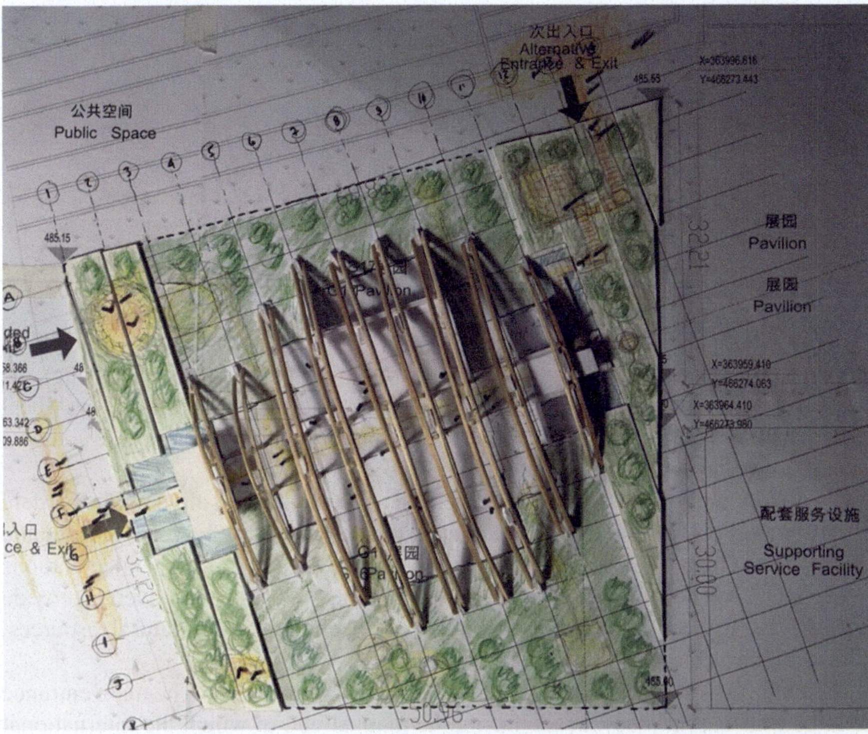

Fig. 3 Study model by Studio Cardenas Conscious Design

poles arches create a welcoming biomorphic vaulted interior space, which embraces the visitors (Fig. 4). From the exterior, the visual impact of the Garden Pavilion is

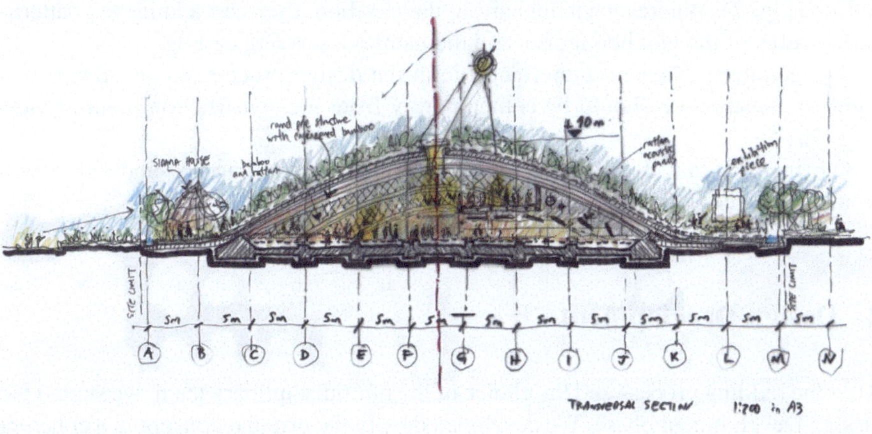

Fig. 4 Sketch by Mauricio Cardenas Laverde, showing concepts of passive design and form finding

Fig. 5 Study model, the visual impact of the Garden Pavilion is that of a hill

that of a hill, the soft green slopes are inviting tourists to visit and to walk through the garden to admire the variety of bamboo species arranged together with several local plants. The geometry of the pavilion is straightforward, and the bamboo structure is rational, modular, and contemporary (Fig. 5) [3].

The use of bamboo for this project, dried for construction and alive for the garden, is strategical as it is a natural resource contributing directly to at least six Sustainable Development Goals (SDGs) adopted by all the member states of the United Nations for the 2030 Agenda for Sustainable Development, such as: end poverty, provide affordable energy, access to affordable construction, efficient use of natural resources, address climate change and protect ecosystems [4].

The functions are under the Garden, under the bamboo arches. The main entrance is a large transparent glass façade facing the main street, of which the International Pavilion is in front. The idea is that to access the INBAR Pavilion is necessary to walk into the garden, passing in between the water ponds, and then going inside. The interior is a spacious main open space, is the central focal point around which all activities are organized, ideal to host exhibitions, presentations, events, etc. (Fig. 6). The main space is filled with natural light, which enters into the spaces in a magical way through the "cuts" in the Garden and is filtered through the bamboo-weaving ceiling (Fig. 7). Wherever visitors are in the pavilion, they can admire the patterns and textures of the bamboo arches and the bamboo-weaving ceiling.

The geometry plays an important role in the design process, as our idea is that bamboo construction should be contemporary from the construction point of view (Fig. 8).

Our intent is that Bamboo Eye will promote the use of bamboo and rattan with an innovative contemporary architectural language.

2 Design Development

After the bidding process and the choice of the multidisciplinary team, we started the Design Development phase. We developed deeply the original concept in a coherent

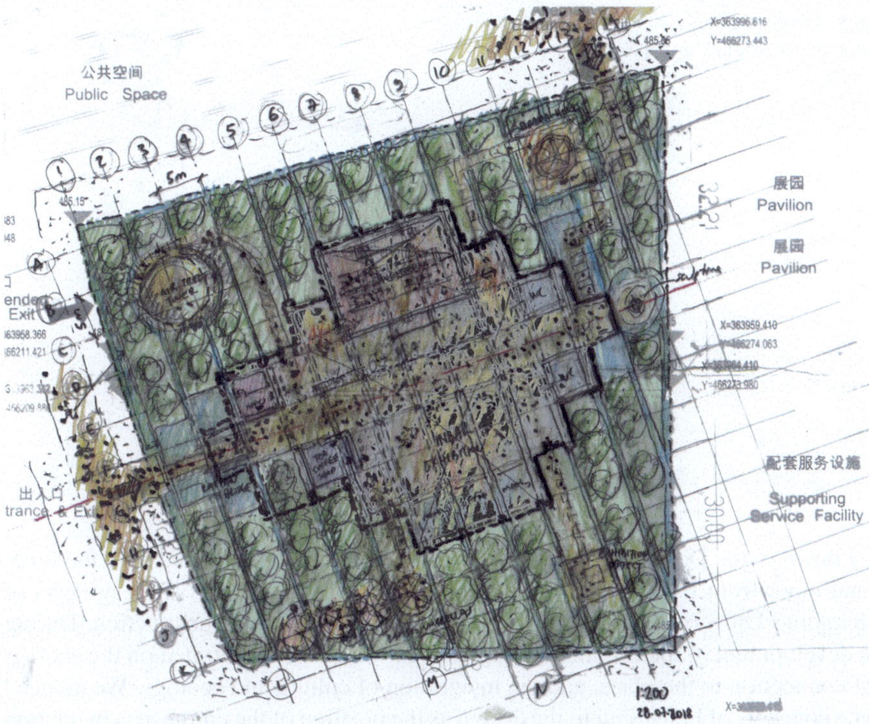

Fig. 6 General plan. Sketch by Mauricio Cardenas Laverde

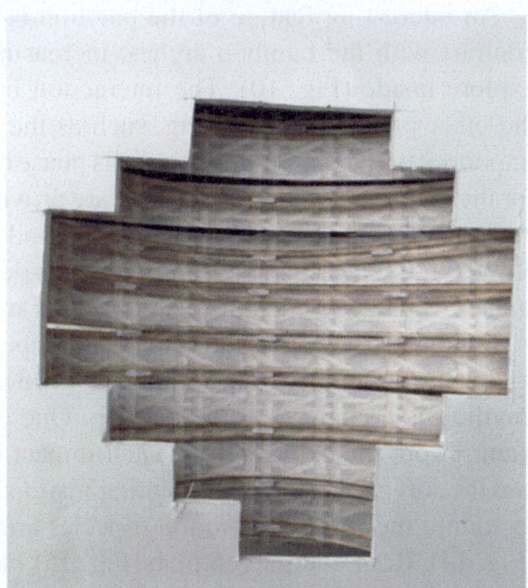

Fig. 7 Study model of the bamboo-weaving ceiling

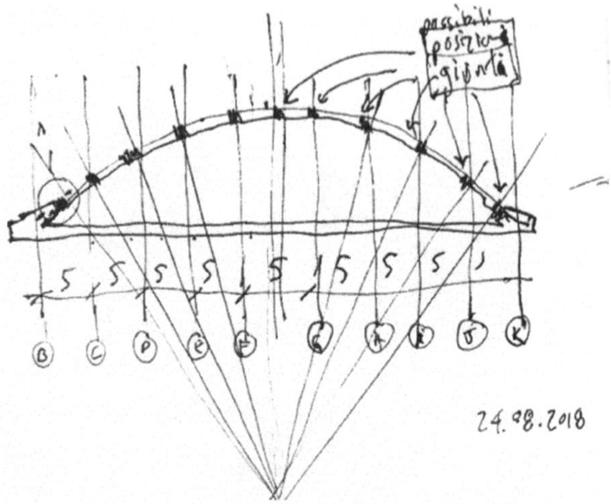

Fig. 8 Geometry study sketch by Mauricio Cardenas Laverde

way down to the Detail Design phase. The architecture was enriched with the functional input from INBAR and the technical input from architects and engineers of Chongqing University, and from Chinese experts on bamboo construction. During the development of the original concept design, we also concentrated on the ecological connection to the place, and the integration of culture and ecology. We induced some concepts of Feng Shui to the design as the position of the entrance is in relation to the Garden and to the free flow of energy through the spaces, as shown in the general plan of the garden (Fig. 9).

An interesting feature of the pavilion is those large glass facades, which are in contrast with the bamboo arches, increasing the curiosity and inviting visitors to explore inside (Fig. 10). The interaction between the natural bamboo arches and the other parts of the building, such as the glass facades and the green roof, were important parts to develop during this phase because an innovative approach was used for the architectural design. Details are provided in Figs. 11 and 12. Usually, bamboo constructions are located in the tropics and therefore do not need air-tight facades. However, the fact that the Bamboo Eye is designed and built with high-performance standards makes it a unique example of a sustainable contemporary building [5].

Following the concept idea that landscape and architecture fuse into one, the green roof is the continuity of the garden (Fig. 13). The Garden embraces the Pavilion, the Pavilion is contained in the Garden. One cannot exist without the other like Yin Yang. From the exterior, the visual impact of the Garden Pavilion is that of a hill, and the soft green slopes are inviting visitors to come and to walk through the garden to admire the variety of bamboo species arranged together with several local plants (Fig. 14). Roof and garden plan (Fig. 15) including the description of all the plants species locally available. Various species of plants mixed with different kinds of local bamboo are included in the Garden. These are located strategically to create "comfort zones" with shadows and specific viewing points to admire the Garden Pavilion.

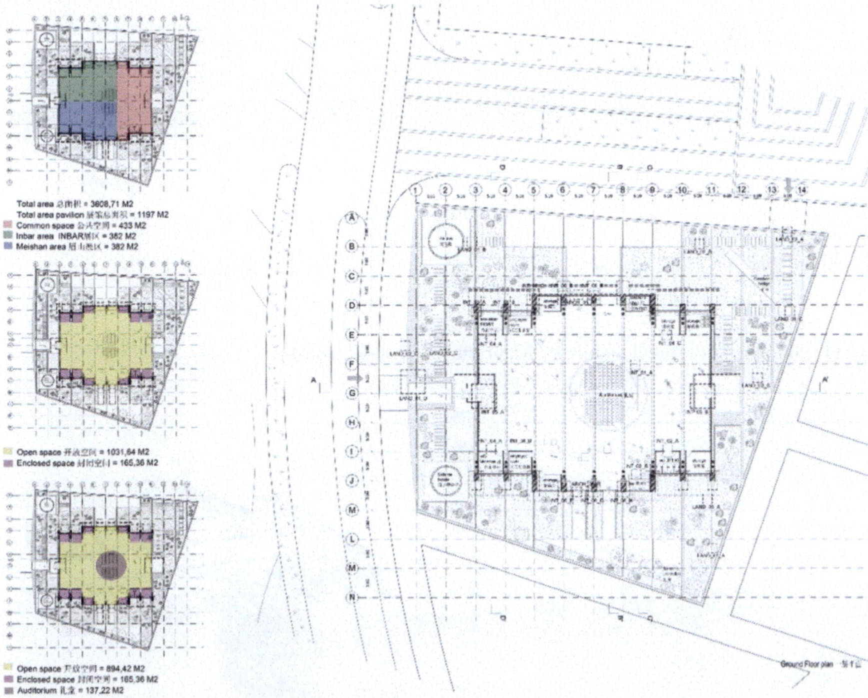

Fig. 9 General plan. Drawing by Studio Cardenas Conscious Design

Fig. 10 Front elevation. 3D model drawing by Studio Cardenas Conscious Design

We used natural materials to imply the local tradition and create new relationships between them. All local materials such as bamboo, water, pebbles, rattan, soil, plants were used in a contemporary way, giving continuity to local tradition towards innovation (Fig. 16). The main colors of the project are natural colors such as dried-bamboo yellow, life bamboo green, different tones of green of plants and yellowish-brown of the bamboo-weaving ceiling. During clear days, the blue sky penetrates inside the pavilion through the skylights.

The form and the orientation of the Pavilion are the results of studies and the application of strategies of passive design to achieve the best use of natural light and

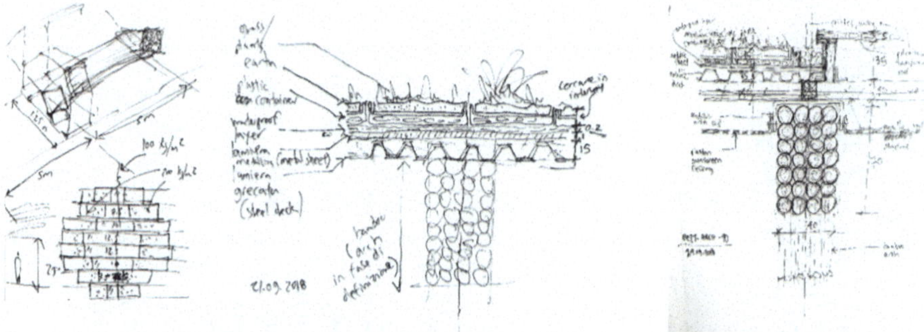

Fig. 11 Sketches of typical details by Mauricio Cardenas Laverde

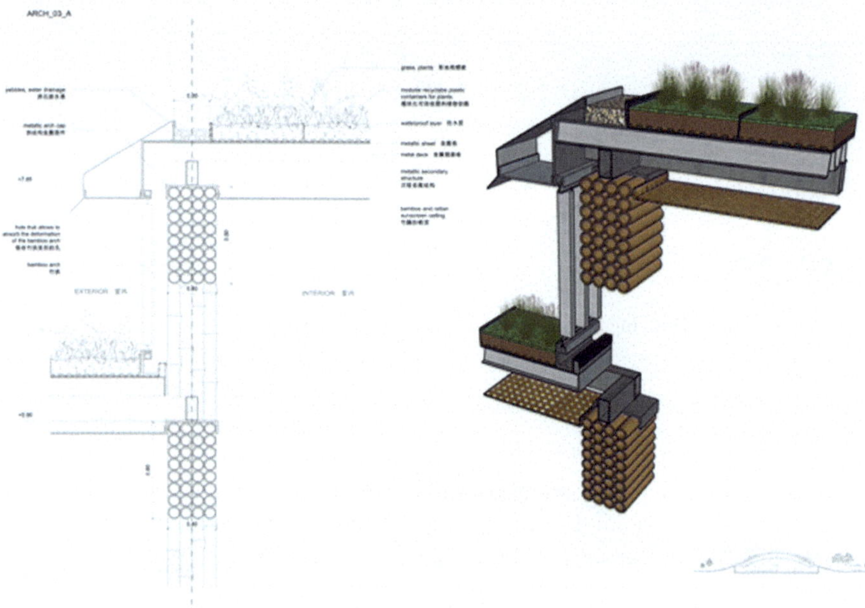

Fig. 12 Typical detail. Drawings by Studio Cardenas Conscious Design

natural ventilation (Fig. 17). The objective is that the breeze will flow freely inside of the interior spaces entering from grilles between the bamboo arches, allowing cross ventilation and reducing the use of air-conditioning. If required, the openable glass facades fins can be actioned to increase the natural airflow. The transparent area of green roof was studied to invite natural light inside during the whole day, so as to limit the use of artificial light when natural light is not available [6].

The geometry of the pavilion is straightforward, is rational, modular, and contemporary. The arrangement of 18 natural bamboo poles for each arch breaks the visual impact of the mass into a series of lines which perceived together with the bamboo

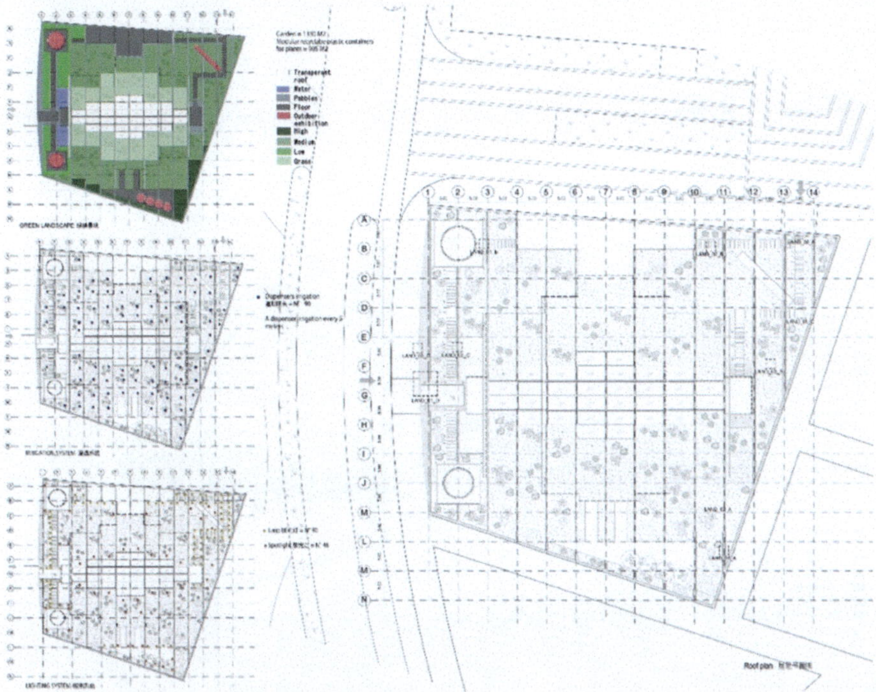

Fig. 13 Garden and roof plan. Drawing by Studio Cardenas Conscious Design

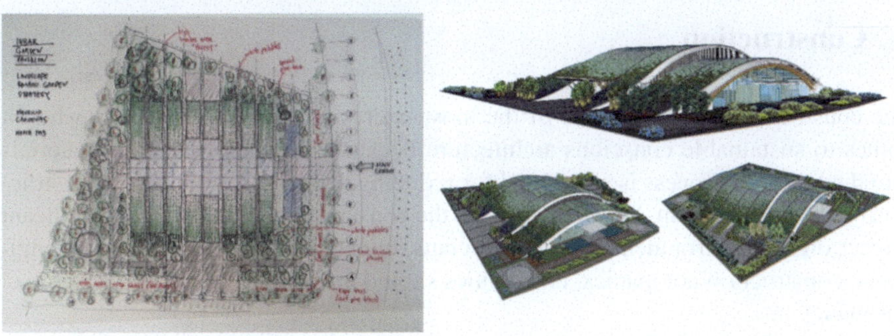

Fig. 14 Studies of the bamboo garden. Sketch by Mauricio Cardenas Laverde, 3D model images by Studio Cardenas Conscious Design

weaving mesh generates a vibrant complexity to the eyes of visitors. The 32 m span bamboo arches are the bearing structure as well as the main architectural feature (Figs. 18 and 19).

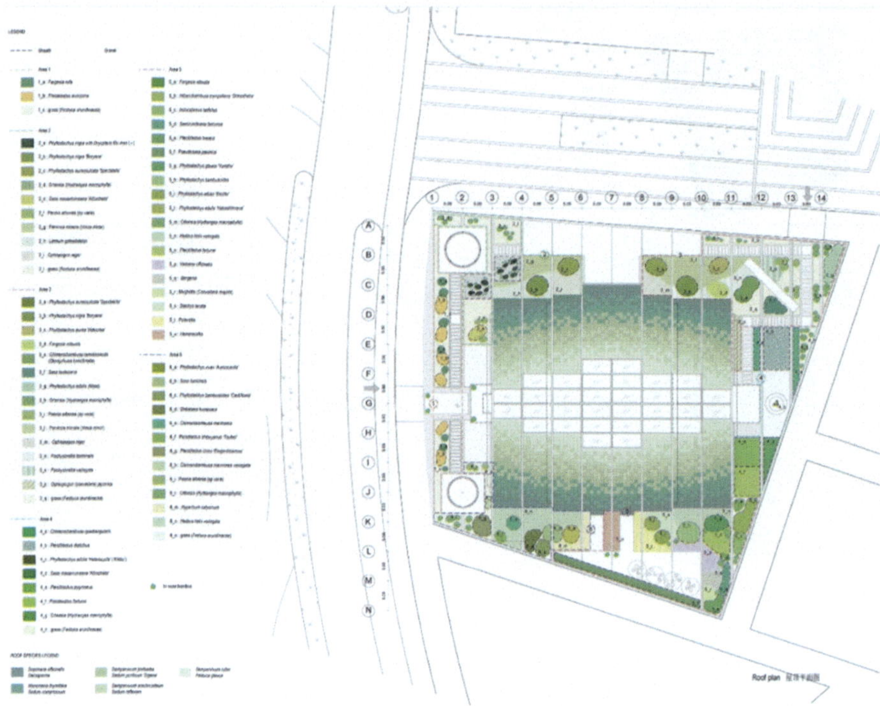

Fig. 15 Roof and garden plan. Drawing by Studio Cardenas Conscious Design

3 Construction

The construction process is one of the most interesting phases, especially when it comes to sustainable conscious architecture because introducing natural materials into the building process is strategical for reducing the negative impact of construction on the environment. It is no secret that the construction industry has a significant impact on the environment, so all politicians, clients, developers, architects, engineers, construction companies, universities should have the responsibility to protect our planet.

For an environmentally friendly construction, it was crucial for the team to work all together starting from the pre-construction stage as described in the Design Development. Regarding the bamboo structure, our design uses treated bamboo culms (Fig. 20) and technologies of bamboo bending (Fig. 21). The architectural design takes the maximum bending of the bamboo into account and architects work together with the engineers during the construction process. 9 Bamboo arches are excellent structural elements, each of which composes of 18 bamboo culms, achieving a 32 m span and a height of 9 m. The pavilion is the largest bamboo structure in Northern China [7]. Each bamboo arch is designed as a truss arch, which takes advantage of not only the arch structure but also the truss structure, making this large-span

Fig. 16 Local materials such as bamboo, water, pebbles, rattan, plants are used in a contemporary way, giving continuity to local tradition towards innovation

bamboo architecture look very light. The pavilion uses more than 5000 *phyllostachys pubescens* bamboo culms, 8 to 10 cm in diameter, with an age between 4 and 6 years (Fig. 22). After a special process, bamboo culms become golden yellow. The service life of this special bamboo extends to 30 years, greatly expanding the applicable scene of bamboo construction.

Thanks to the cooperation with the engineering team of Chongqing University (Fig. 23), it was possible to reduce excavation and use concrete only for those plinths on top of which the bamboo arches lean. The green roof has no concrete too, with the humid soil and plants to give weight, stability to the bamboo bearing structure as well as provide a great acoustic buffer and effective temperature control. The green roof is the continuity of the garden, planted with sedum, sempervivum, delosperma, saponaria, micromeria thymifolia and etc. (Fig. 24). A subtle contrast is created by those green bamboo plants at the Garden and dried bamboo poles of arches in the Pavilion.

The ceiling is another example of the use of bamboo and great Chinese handcraftsmanship (Fig. 25). The idea is to apply traditional Chinese handicrafts to large-scale buildings, providing a warm familiar and welcoming environment for visitors.

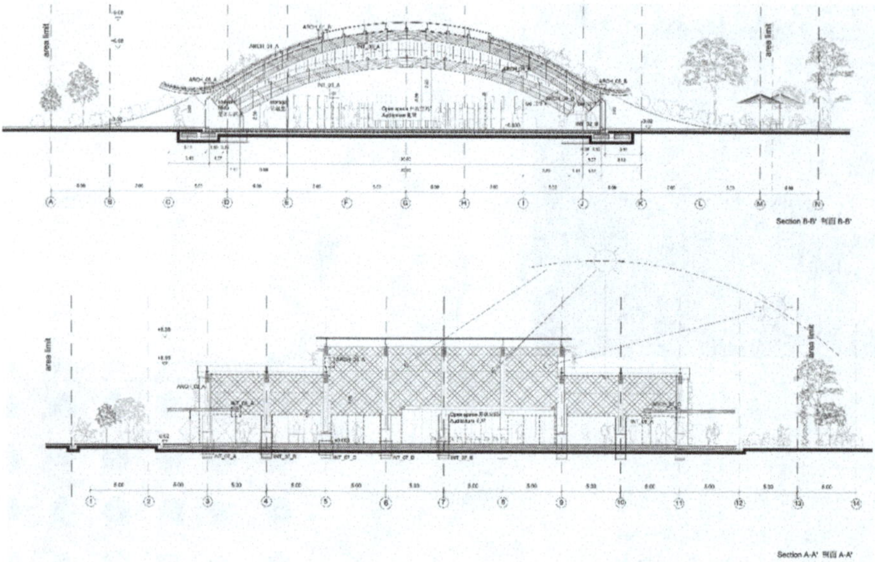

Fig. 17 Longitudinal and transversal sections. Drawings by Studio Cardenas Conscious Design

Fig. 18 The strong symbiosis between architecture and landscape is achieved aesthetically through soft shapes and forms. Renderings by Studio Cardenas Conscious Design

Natural light shines through the bamboo top onto the bamboo ceiling, reflecting beautiful colors (Fig. 26). Shadows are clear, soft, and vibrant (Figs. 27 and 28).

Wherever visitors are in the INBAR Garden, they can admire the patterns and textures of the bamboo arches and the bamboo weaving, feel the constant breeze and the fragrance of the bamboo poles, the warmth of the direct sunlight rays filtering through the bamboo weaving ceiling, as shown from Figs. 26, 27 and 28.

Fig. 19 Aerial view of the pavilion. Renderings by Studio Cardenas Conscious Design

Fig. 20 Treatment of bamboo poles in Meishan, Sichuan. Photos by China Bamboo Company

INBAR Garden Pavilion received a Grand Award in the category of International Outdoor Garden in recognition of its innovative and ecological design and execution [8].

Fig. 21 Bending bamboo culms for large span arches. Photos by Shidong Nie

Fig. 22 Natural bamboo poles were used for the arches. Photos by Chongqing University

Fig. 23 Team work. Photos by China Bamboo Company

4 Experience

Since the first day, the experience was very exciting! It all started on a hot morning in July 2018, around 11 am Milan time, I received an unexpected phone call from Ms. Liu Kewei, Coordinator of Global Bamboo Construction Programme of INBAR, inviting me to participate in the competition for the design of the INBAR Pavilion. Was such a pleasant surprise and honor!

Fig. 24 Green roof under construction. Photos by Mauricio Cardenas Laverde

Fig. 25 Ceiling of Bamboo Eye under construction. Photos by Mauricio Cardenas Laverde

Fig. 26 Views of the bamboo arches and bamboo weaving ceiling. Photos by Pan Zhenyu

Fig. 27 Interior view of Bamboo Eye with its exhibition. Photo by Pan Zhenyu

Fig. 28 Vibrant bamboo arches and bamboo weaving ceiling. Photo by Pan Zhenyu

Few days after winning the competition I traveled to Beijing to work directly with INBAR on the development of the design and to meet the candidates for the role of EPC (Engineering Procurement Construction). Create a highly professional team is crucial to achieve a high standard design and construction that INBAR is very clear. The General Research Institute of Architecture and Planning Design of Chongqing University was the selected EPC, with which I started to cooperate from the pre-construction stage as described in the Design Development. I moved

Fig. 29 Aerial view of INBAR Pavilion. Photo by Pan Zhenyu

Fig. 30 Aerial view, with mountains in Yanqing at the background. Photo by Pan Zhenyu

Fig. 31 Garden and Pavilion melt into one. Photo by Pan Zhenyu

Fig. 32 West façade of INBAR Pavilion. Photo by Pan Zhenyu

Fig. 33 The name of INBAR Pavilion, Bamboo Eye, comes from the reflection on the water. Photo by Pan Zhenyu

Fig. 34 Various species of plants mixed with different kinds of local bamboo in the Garden. Photo by Pan Zhenyu

to Chongqing where we worked very closely, also with China Bamboo Company, developing the design in all its parts and details following the original Concept Design. The experience was very fulfilling, the exchange of knowledge, information and coordination between the team in Milan, myself and the EPC in Chongqing and INBAR in Beijing worked very well.

After the precedent intense work last two months, the team was ready to deal with the preparation for the construction. Part of the process I was back in Milan but we kept fluent communication. During the second half of the construction, I was back to China. Whenever we encounter problems, we met to discuss in Chongqing, in Beijing, or at the construction site. For any challenge we encountered we always had the support from INBAR.

Fig. 35 Consistency and coherence during the conscious design process to the construction

In conclusion, the experience as a whole was better than expected. Of course, there were difficult moments, difficult problems to solve, but the key of success of the design and built was team work. We made the consistency and coherence during the conscious design process to the completed construction (Fig. 35).

5 Photos

Several photos related to different phases of this project are provided as below, as shown from Figs. 36, 37, 38 and 39.

Fig. 36 Project Design Development presentation to Co-Chair of INBAR Board of Trustees, Prof. Jiang Zehui, in Beijing, China in August 2018

Fig. 37 Project Design Development presentation to INBAR Director General, Dr. Hans Friederich, in Beijing, China in August 2018

Fig. 38 Design and construction team photo during construction, March 2019

Fig. 39 Several team members attending INBAR Honour Day, May 2019

6 Profile of the Project

Completion Date April 2019 **Location** 2019 Beijing International Horticultural Exposition, Yanqing District, Beijing, China **Client** INBAR—International Bamboo and Rattan Organization **Construction Area** Total area: 3600 m^2 Total garden area: 2000 m^2 Total interior area Pavilion: 1600 m^2 **Concept** Studio Cardenas Conscious Design: Arch. Mauricio Cardenas Laverde **Architecture Design** Studio Cardenas Conscious Design: Arch. Mauricio Cardenas Laverde Arch. Monica Bernardi Arch. Chen Liyuan Designer Roberta Zecca General Research Institute of Architecture and Planning Design Co., Ltd of Chongqing University: Prof. Arch. Wang Xuesong Prof. Arch. Wang Zhaoxia Prof. Arch. Qin Lin	**Landscape Design** Studio Cardenas Conscious Design: Arch. Mauricio Cardenas Laverde Arch. Monica Bernardi Arch. Chen Liyuan Designer Roberta Zecca Lorenzo Bar—Milan, Italy Wang Xian, Wang Kailin—Beijing, China **Construction** EPC (Engineering Procurement Construction) by General Research Institute of Architecture and Planning Design Co., Ltd of Chongqing University: Prof. Eng. Cui Jia Prof. Eng. Nie Shidong Prof. Arch. Wang Xuesong Prof. Arch. Wang Zhaoxia Prof. Arch. Qin Lin **Construction Supervision** Studio Cardenas Conscious Design: Arch. Mauricio Cardenas Laverde General Research Institute of Architecture and Planning Design Co.Ltd of Chongqing University: Prof. Arch. Wang Xuesong Prof. Arch. Wang Zhaoxia Prof. Arch. Qin Lin

References

1. Cardenas, Laverde M. 2013. Environmental comfort, 718–739. The Architect's Almanac, ed. by Piano, R. Bologna, Italy: Proctor Edizoni S.p.A.
2. Cardenas, Laverde M. 2008. Bamboo as a construction material. Napoli, Italy: Editorial Group Esselibri—Simone.
3. Cardenas, Laverde M. 2020. Building with bamboo, 135–139, 144–145. In Architecture of after building with plants, eds. Corrado, M. Rome, Italy: Derive Approdi srl.
4. Cárdenas, Laverde M. 2011. Preface, 6–11. *Bamboo*, by Sánchez Vidiella, A., Frechmann Kolón. Köln, Germany: GmbH.
5. INBAR homepage. https://www.inbar.int/bamboo-eye-pavilion/. Last accessed 19 July 2021.
6. Cardenas, Laverde M. 2020. The conscious city. In *Catalogue of the Italian design day 2020, drawing the future*, 42–43, Milano, Italy: Fondazione ADI.
7. INBAR homepage. https://www.inbar.int/largest-bamboo-structure-in-north-china-set-to-open-in-expo-2019. Last accessed 19 July 2021.
8. INBAR homepage. https://www.inbar.int/inbar-pavilion-wins-grand-award. Last accessed 19 July 2021.

House Loti

Audrey Mertens, Chen Qi, and Fantine Fontaine

Abstract Designed to promote and demonstrate the utility of bamboo in architecture and engineering, House Loti takes the form of a blossoming lotus. The authors came together to propose a pavilion that takes inspiration from traditional bamboo, and rattan craftsmanship. We sought to deliver a project which is utilitarian, adaptable, and able to be rapidly assembled. The final project was assembled by 4 people under 24 man-hours, installed on-site at INBAR Pavilion at the 2019 Beijing Horticultural Expo.

Keywords Bamboo · House Loti · Pavilion

1 Design Concept

1.1 Context and Brief

For the 2019 International Bamboo Construction Competition, the building was to be designed as a singular pavilion within a 3 m by 3 m footprint, under 6 m tall. The brief asked for shelter which can be adapted for different functions including the ability to rest.

Expanding on the objective goal of the project, there also exists a strong theme that we must demonstrate. The project should utilize bamboo primarily, and given the nature of the International Bamboo and Rattan Organisation (INBAR), the building should sustain a strong overarching concept which experiments with and celebrates the properties of bamboo.

A. Mertens (✉) · F. Fontaine
Urban and Environmental Engineering Department, University of Liège, Liège, Belgium

C. Qi
Faculty of Environment and Technology, University of the West of England, Bristol, UK

K. Liu et al. (eds.), *2019 International Bamboo Construction Competition*, Springer Tracts in Civil Engineering, https://doi.org/10.1007/978-3-030-91990-0_5

Fig. 1 3D model of the pavilion as designed

1.2 Spatial Design

Given the unique nature of a pavilion, our spatial response to the brief was one of independence. The building is in essence without context, able to be temporarily or permanently installed anywhere. As such, our building would have a ground floor area that is welcoming when approached from any direction. Further, emphasized with the modular collapsible "petals" of the façade, the proposed project allows the user to adjust their exposure to the environment (Fig. 1).

The stand the team is taking here is to promote creative typologies, which is the way to put bamboo on the map as a construction material for other buildings in the future.

Contrasting against the ground floor, a smaller second-level area is designed to create a sense of privacy. Accessed via a ladder, the second level exists in a dichotomy to the ground floor; the area is enclosed with a fixed building envelope, and a series of traditional rattan weave provides a shelter from light and heat.

1.3 Structure and Construction

The construction response captures two main purposes of the building, primarily it should fully utilize the unique properties with bamboo while secondarily the structure should be easily constructible on site.

The main framework of the building sees the use of six bundles of natural bamboo, beginning at the foundation and meeting at their apex. The tensile property of the bamboo allows the building to sit in constant tension, such tension would resist any variable horizontal load including wind or people standing against it.

This hexagonal symmetry allows six equal sections of the building to be prefabricated. As seen in Fig. 2, each symmetrical section of the building consists of a

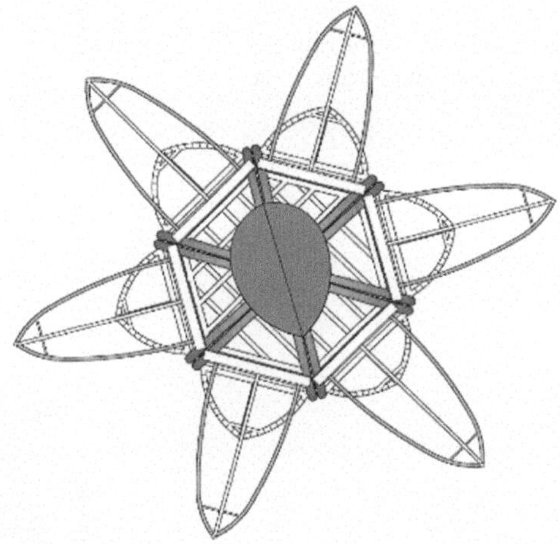

Fig. 2 Layout plan of the design

self-contained module with two vertical main bamboo columns and two horizontal bracing components.

The entire structure sits above 6 foundational concrete blocks cast in-situ. Each of these blocks hosts a vertical steel tube to which the vertical bamboo columns will attach to. This allows the building to be quickly installed with minimal tools.

1.4 Material

The exploded axonometric drawing in Fig. 3 displays the detailed strategy. Each panel can be broken down into 3 main materials: bamboo, rattan, and fabric.

Bamboo is used in the project as the main load-bearing component. Bamboo experiences both compressional and tensile stress in the pavilion. Bamboo also allows for circular incisions to be made to allow for either friction joints or the creation of movable hinges such as the component used in the "petals" of the building.

Rattan functions as the secondary structure and as a layer of cladding on the building. This provides marginal bracing for the bamboo and shelters the building from the sun. It also acts as a secondary structure to which the weatherproof fabric layer would be attached.

Waterproof Fabric is used to provide shelter from the rain, sun, exposure, and moisture. As prescribed by Kaminski et al. [1], bamboo should not be exposed to water nor rain. Thus, this fabric protects the structure from direct rain, to prevent the bamboo from rotting, poles must be treated before implementation.

Attached to the exterior of the building, combined with the rattan weave provides a lightweight cladding solution for the pavilion. It is also easily repairable using spare material with primitive tools. However, due to the unavailability of a fit fabric at the time, the entrepreneur in charge of prefabrication of the panels replaced it

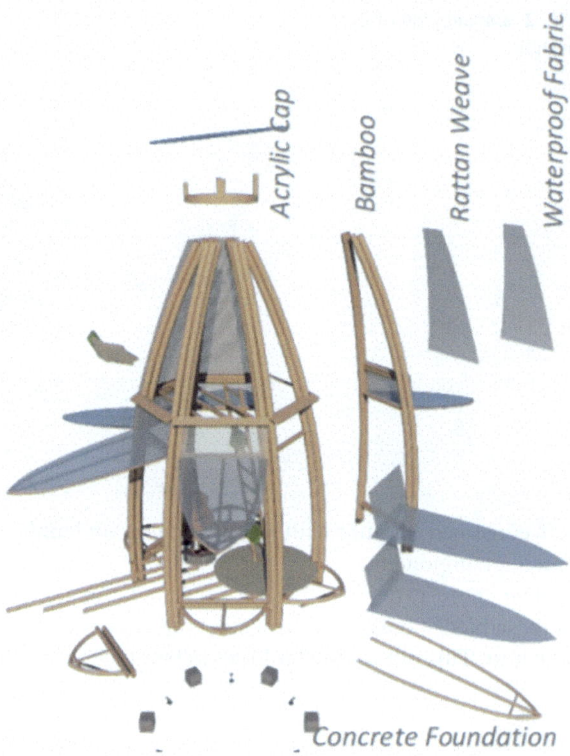

Fig. 3 Exploded axonometric scheme of the layers and components of the structure

with a Plexiglas layer, therefore still protecting the bamboo structure from rain and maintaining the design will keep partial transparency of the wall cladding.

1.5 Kinetic Facade

The ground level of the building is sheltered by 6 symmetrical "petals", each attached to a module of the building. A steel cable attaches through a pulley system to the mid-point of each petal, the 6 cables are threaded through a secondary pulley at the pinnacle of the structure, down through a stabilizing component as shown in Fig. 4. These cables are wound through a hand crank in the interior of the building, where it can be operated by a single person. This mechanism allows the petals to be opened, closed, or secured at any position in between.

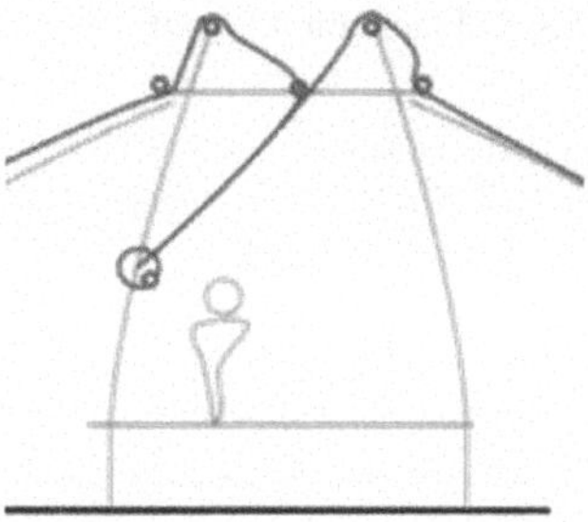

Fig. 4 Principle diagram for the cables allowing to lift the "petals"

1.6 Site Strategy

While the current project is designed for a pavilion setting, where a single installation is created, we foresee possibilities for wider implementation of the design. One potential use of these pavilions is at designated camping areas in national parks or natural resorts. This allows a variable amount to be places, adaptable to a range of different densities. While the current construction is not suitable for more than a couple of overnight stays, we are confident it can be adapted and improved to provide more permanent shelter when required.

2 Design Development

2.1 Foundation Design

For the 2019 International Bamboo Construction Competition, the building was to be designed as a singular pavilion within an area of 9 m^2 (3 m by 3 m), as well as a projection area no larger than 16 m^2 (4 m by 4 m), including cantilever structural elements such as canopy.

As previously explained, bamboo should not be exposed to water nor the humidity from the soil [1]. To prevent the bamboo from rotting, foundations are made of concrete.

Thus, to be as economical as possible in non-renewable materials such as concrete and steel, the foundations of the House Loti were kept to a minimum of 6 concrete studs, each one topped by steel anchors to bolt the bamboo pole to, positioned in a 3 m wide hexagon. As shown in Figs. 5 and 6 small additional concrete studs serve as holders for the seating's legs.

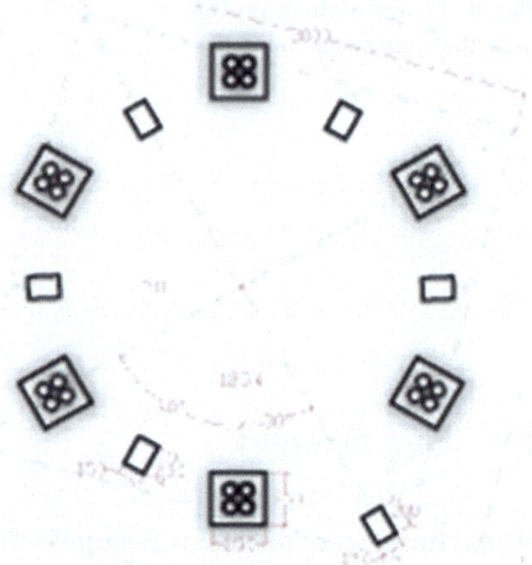

Fig. 5 Foundation layout

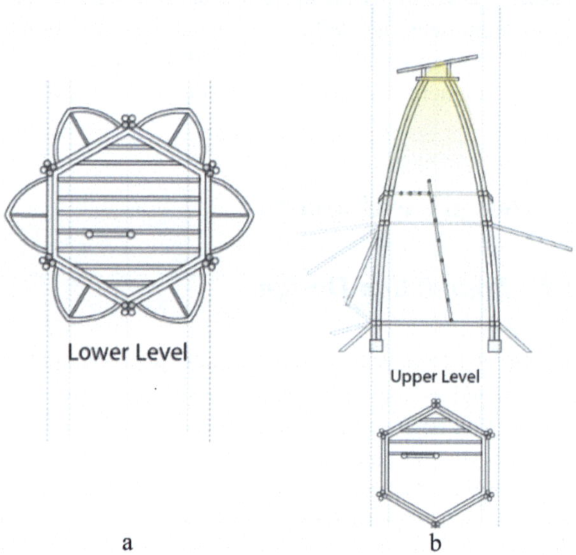

Fig. 6 Layouts and cross section of the pavilion: Lower level (**a**); upper level (**b**)

2.2 Ground Floor Design

The ground floor consisted of a hexagonal main deck, with laminated bamboo flooring. Rotatable bamboo petal-shaped seating can unfold and create extra flooring or a nice stop to sit on, accessible from the exterior of the structure. These seating have a small bamboo pole used as a leg to support the deck. These supports fit in a notch, provided for that purpose in the concrete studs below. The entry of the pavilion lies in one of these petals, which is legless, and which contacts the ground directly and creates a small ramp to climb onto the main deck of the building. Unfolded, the floor plan recalls the shape of a flower.

This floor is as polyvalent as can be: the deck is vast, regarding the mission statement's constraints, creating a comfy little reunion spot, sheltered from the rain whilst still well ventilated and open to the exterior for sunnier days. It can also be used as 5 different meeting spots for couples meeting on different petals, taking a break, tea, or lunch. If closed, the lotus can also be a nice place for self-examination, to meditate or just to chill. The floor plan is shown in Fig. 6a.

2.3 Second Floor Design

The second floor is a small mezzanine bamboo floor, taking up just half of the hexagonal shape of the building, thus leaving a full height and view of the ceiling for half of the ground floor. It is accessible by a steady ladder, allowing one person at a time to come up and enjoy this small bamboo nest. This upper floor plan is shown in Fig. 6b.

A small basket, hanging on a pulley, allows you to conveniently carry your belongings up with you, without having to carry them whilst going up the ladder. It is a nice spot to get a glance at the view, taking a bit of height. It is a place to store stuff or to rest a little by yourself.

2.4 Ceiling Design

A round sky-opening connects the 6 panels. This hole allows the air to flow through the building and up. A small plexiglass leaf-shaped screen is mounted on the very top of the structure, completing its uniquely natural look, and sheltering the volume of the building from the rain.

2.5 Facade Design

The entire shape of the building evokes the delicacy of a lotus flower. Arches not only highlight the flexibility characteristics of bamboo; it is also very soothing for the eye. The facade features 6 kinetic elements as mentioned before and creates a distinct appearance and flexibility of function for the building. Figure 7 demonstrates the profile of the building when the kinetic elements are opened and closed. A photographic depiction of the facade can be found in Sect. 4.

The transparency of the structure allows a lot of light in. The rattan weave not only gives the building a unique look but it also values local craftsmanship and provides a visual cladding from a distance into the building whilst still allowing full

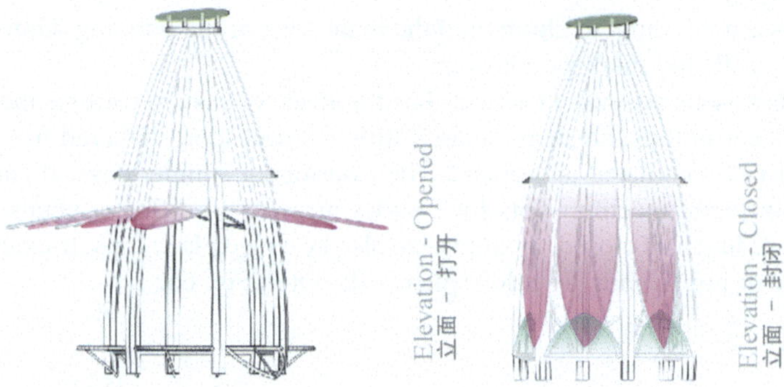

Fig. 7 Elevations of the building

vision out. The kinetic elements allow us to adjust the level of privacy, exposure, weatherproofing, circulation, and footprint of the building, making it suitable for a wide range of locales and functions.

2.6 *Structural Approach*

The calculation of the structure was guided by methods taught at the firm ASALI BALI during their bamboo construction workshops and by Kaminski et al. writings [2–4]. The calculation is based on the ISO standards and Eurocodes [5–8].

To check the deformation of the pavilion, the structure was modelled with Rhino 3D and the Grasshopper plugin. Then, it was transferred to Autodesk software ROBOTO, entering Moso bamboo's properties manually for simple tubular beams. The results verified the resistance of the structure for all elements in compression, tension, and bending and gave a maximal displacement at the top of the structure equal to 5.9 cm. This displacement, corresponding to H/93, was considered acceptable for a temporary building.

The authors want to highlight the fact that the design was several times presented to local Balinese entrepreneurs and the architect Made Wirahadi Purnawan, to get iterative feedback on versions of the design. This expert advice, together with the knowledge brought up by craftsmen working daily with bamboo, its forces, and constrains, really enriched this design process way more than any engineering calculation could have.

3 Construction

3.1 Pre-fabrication

The initial step of construction happens off-site. We sent the contractor a Sketch Up module containing all bamboo components of the building. The building is broken down and manufactured as independent components, then assembled in the factory to create 6-panel modules. The detailed construction is as follows, as shown in Fig. 8.

3.2 Module Delivery

The structure is made of six frames that are assembled with rope on-site into a hexagonal shaped plan, primary floor structure, and end cap on top. Each frame was prefabricated by the contractor off-site and brought on-site by truck. The frames can be carried by hand by 3 men.

3.3 Foundation Casting

The foundation is cast on-site as semi-submerged concrete blocks (Fig. 9a). During casting, stainless-steel tubes are inserted into the foundation. This will allow main structural bamboo columns to be inserted onto the foundation. Elevation of the base of the bamboo above ground to prevent weathering damage.

3.4 Module Preparation

During transit, the prefabricated panels (Fig. 9b) are braced with additional elements. These are bamboo strips running perpendicular to the modules, nailed in place. It also secures the kinetic facade in place. Prior to the installation, we will extract these bracing elements and test the hinges of the petals.

3.5 Initial Module Installation

To begin the installation, we erect a single module with 3 people, inserting the bamboo into the foundation cast (Fig. 10). Ensuring that both sides of the panel rest fully on the foundation, a single bamboo pole would be used to support this module as we begin to work on the other modules.

a

Step 1: Cylindrical bamboo is cut to size as needed in the building. The primary beams are pre-stressed to create a slight bend.

b

Step 2: Cap is created. A single piece of die-cut stainless steel is bent to create a crown. A leaf-shaped acrylic element is laser-cut.

c

Step 3: Separate from the main structure, the petals are created. Rattan is attached to the surface of panels. An acrylic sheet is used for the exterior instead of fabric, that was a compromise we agreed to with the contractor due to a lack of material.

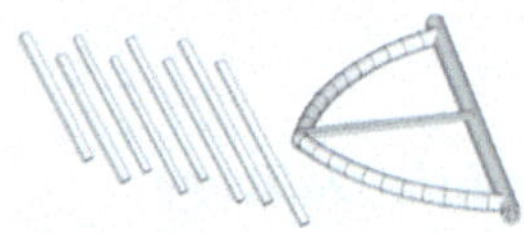

d

Step 4: Adjustable steps are created separate from the main building. Pre-stressed bamboo forms the frame while bamboo sliced in half vertically is installed as the decking.

e

Step 5: The main module frames are created out of bamboo. The petals and adjustable steps are inserted into slots on the main bamboo columns.

f

Step 6: Rattan is laid onto the upper section of the frame and secured in place. A layer of acrylic is added above.

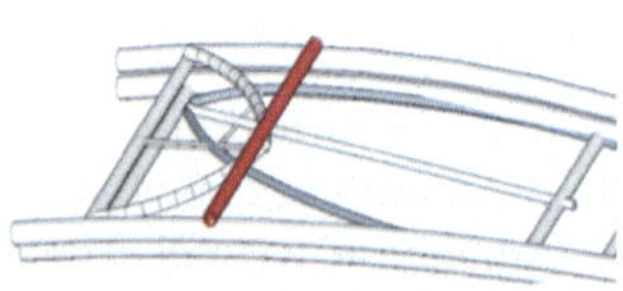

g

Step 7: Temporary bracing is added to the structure to allow the petals to stay in place during transport to the competition ground.

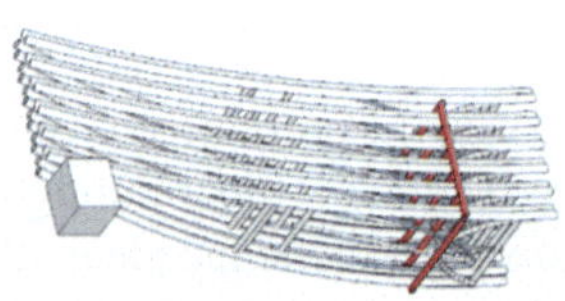

h

Step 8: The modules are bonded to each other using additional temporary braces. This is loaded onto a flatbed truck with additional support.

Fig. 8 Steps for construction

a b

Fig. 9 Casted foundations (**a**); prefabricated panels (**b**)

a b

Fig. 10 Module preparation (**a**); initial module installation (**b**)

3.6 Adjacent Module Installation

The panels are installed one by one. During this phase of construction, we do not secure any module to its adjacent counterpart (Fig. 11a). This demonstrates the advantage of the tensile module frame, as each module can support itself independently of others.

a b

Fig. 11 Adjacent module installation (a); complete the modules (b)

a b

Fig. 12 Adjacent module connection (a); installation of ceiling element (b)

3.7 Complete the Modules

We install each of the modules onto their foundation. We use grass ropes to temporarily secure the components to each other, marking out positions to install bolts. These ropes are later removed once the bolts are installed (Fig. 11b).

3.8 Adjacent Module Connection

Bolts are installed to secure each module to one another (Fig. 12a). Whereas the timber columns provide vertical load-bearing, these bolts provide bracing against any variable lateral load including wind, or any person leaning against the building.

3.9 Installation of Ceiling Element

The roof element includes a prefabricated module of a stainless-steel crown, capped by a slanted acrylic panel (Fig. 12b). This provides resistance against rain while allowing air circulation between the interior and exterior of the building.

3.10 Threading Cable

From the middle of the petals, we secured steel cables that ran through the pinnacle of the pavilion. A series of pulleys provides a point of support. These 6 cables join into a singular cable (Fig. 13a).

a b

Fig. 13 Threading cable (**a**); connecting cables to winch (**b**)

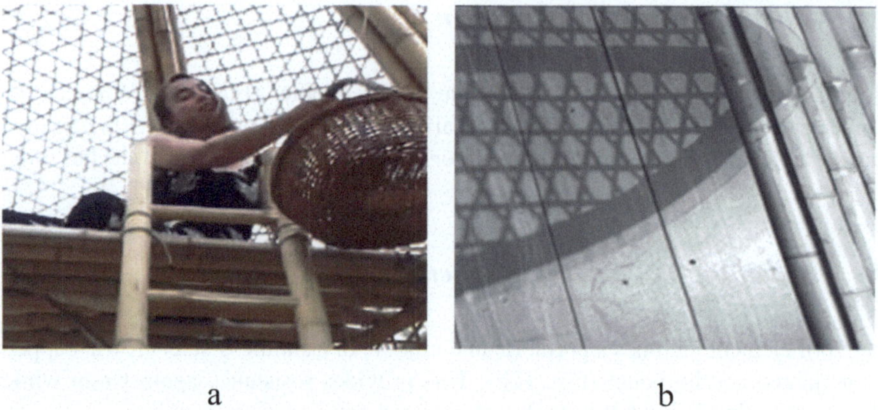

Fig. 14 Upper-level flooring (**a**); lower-level flooring (**b**)

3.11 Connecting Cables to Winch

The singular cable is threaded through a winch at one of the columns on the interior of the building. This winch has a self-locking mechanism to allow analog adjustments to the angle of the petals (Fig. 13b).

3.12 Upper-Level Flooring

The upper-level floor (Fig. 14a) consists of a series of cylindrical bamboo decking. This provides a rigid platform as well as additional lateral bracing for the building. A bamboo ladder is placed to allow access to this level. As well as a pulley to retrieve items.

3.13 Lower-Level Flooring

The lower-level floor (Fig. 14b) consists of cross-laminated bamboo beams. This provides a rigid platform, enough to support a group of people. A small gap is left to allow for thermal expansion and contraction. It also provides additional lateral bracing.

Fig. 15 Final look of the constructed pavilion

4 Photos

The final look of the pavilion is shown in Fig. 15.

5 Discussion

This construction was a huge event in all three of the authors' careers as it was the very first time, we had to handle alone all of the constraints of the design from the very beginning, programming to conducting the worksite and construction on a 1:1 scale. The 2019 International Bamboo Construction Competition was a unique opportunity for us to grow and learn from an actual experience focussing on the constraints and forces that a material has to offer. We were the only team amongst the finalist to take bamboo's bending properties not as a weakness but as a force bringing the curves and shapes of natural bamboo into the design of the structure.

The entrepreneur in charge of building the structure had understood that and the cooperative relationship brought us to eventually win the competition [9]. The original and organic shape was mentioned as a key factor in the jury's final choice, as well as the fact that we had taken advantage of the natural raw bamboo poles in the design.

Again, we want to highlight the fact that close interaction and collaboration between bamboo craftsmen and designers is essential in the development of these organic typologies of buildings.

References

1. Kaminski, S., A. Lawrence, and D. Trujillo. 2016. Structural use of bamboo. Part 1: Introduction to bamboo. *The Structural Engineer* 94 (8): 40–43.
2. Kaminski, S., A. Lawrence, D. Trujillo, and C. King. 2016. Structural use of bamboo. Part 2: Durability and preservation. *The Structural Engineer* 94 (10): 38–43.
3. Kaminski, S., A. Lawrence, D. Trujillo, I. Feltham, and L.F. López. 2016. Structural use of bamboo. Part 3: Design values. *Structural Engineer* 94 (12): 42–45.
4. Kaminski, S., A. Lawrence, D. Trujillo, I. Feltham, L.F. López. 2017. Structural use of bamboo. Part 1: Element design equations. *Structural Engineer* 95 (3): 24–27.
5. European Committee for Standardization. 1995. EN: 384:1995 Structural timber. Determination of characteristic values of mechanical properties and density.
6. European Committee for Standardization. 1995. NF EN 1995—Eurocode 5—Design of timber structures Part 1-1: General-Common rules and rules for buildings.
7. International Organization for Standardization. 2004. ISO 22156:2004 Bamboo—Structural design.
8. International Organization for Standardization. 2004. ISO221557-1:2005 Bamboo—Determination of physical and mechanical properties.
9. INBAR. Full results of the International Bamboo Construction Competition. Retrieved from https://www.inbar.int/full-results-of-the-international-bamboo-construction-competition.

In-Box House

Lelissa Erkissa, Mikyas Tekle, and Simon Mucheye

Abstract As the focus is drawn towards more sustainable construction practices, the use of bamboo as a structural building material is growing as a topic of interest. Composite lumber made from bamboo, termed laminated bamboo lumber (LBL) or bamboo scrimber (BS), has gained the particular interest of researchers and practitioners. In-Box House is a multifunctional bamboo garden house designed to be constructed in a combination of bamboo poles with prefabricated laminated bamboo lumber. The house is to be constructed in three days with four people. The prefabricated modular bamboo lumber with simple connection details make it easy to be constructed on-site. The prefab modular boxes are then inserted into the bamboo structural frames in many flexible multifunctional layouts according to the functional purposes. The idea is that the house can be easily adjusted to any weather type, mood or occasion.

Keywords Inbox house · Laminated bamboo lumber · Multifunctional house · Flexibile architecture · Prefab house

1 Concept Design

1.1 Local Context

Bamboo is a native forest resource in the country Ethiopia. It has over 650,000 hectares of native lowland bamboo (*Oxytenanthera abyssinica*) and over 350, 000 hectares of highland bamboo (*Arundinaria alpina K. Schumach*), the latter representing 86% of Africa's highland bamboo resources. Because of the shortage of proper woody plants for construction in the lowlands, the lowland bamboo is commonly used as an alternative for timber in house construction, for fences, and also as fodder for cattle, as food for people, and as a source of biomass energy [1].

L. Erkissa (✉) · M. Tekle · S. Mucheye
Department of Architecture, Addis Ababa University, Ethiopian Institute of Architecture, Building Construction and City Development (EiABC), Addis Ababa, Ethiopia

K. Liu et al. (eds.), *2019 International Bamboo Construction Competition*, Springer Tracts in Civil Engineering, https://doi.org/10.1007/978-3-030-91990-0_6

1.2 Issue/Problem

Despite having enormous bamboo resources in the region, Ethiopia is now suffering from pressing housing issues resulted from the urban sprawl and rural–urban migration trend, which offers a great example to study the balance between density and quality in social dwelling.

Currently, it is estimated that 80% of the population of Addis Ababa capital city of Ethiopia is living in self-built "slums". One of the most crucial problems in Addis Ababa is the unmet demand for affordable housing. An up-to-95% humanity increase led to rapid urbanization or even over-urbanization, which bring the city to poverty and unemployment. Meanwhile, due to the rare supply of housing by the government, the housing shortage directly resulted in the self-built informal settlements—slums. As the country's recent Millennium Development Goals needs assessment study indicates, the proportion of the urban population living in slums is as high as 80% [2].

The issue of informal settlements involves a wide range of subjects including social, economic, spatial, legal, and political aspects. These aspects in relation to a given context create phenomena that are very complex. The unique social context also provides a wide range of possibilities in the definition of urban life, making it a real challenge for architects to take positions. Setting the boundary of a research project of a complex subject with little locally based background material is challenging.

1.3 Design Program

Humans are constantly reimagining and redefining what a house means according to their needs. The definition of home in Addis Ababa has the more different conditions when compared to the conventional definition of home. Home in Addis Ababa is not only for a living space for a certain family; rather it is a combination of living and production spaces, even for commercial activities. For the residents, a home is a place where they not only live, eat and sleep, but also find a way to generate an income by using the space for production and business. Because of this unique way of living, there is a huge practice and demand for the extension and integration of spaces.

1.4 Proposal

In-Box House: In this particular project, we try to blur this gap between the residents and the architect, by seeing the situation from their perspective of home. We propose a solution that uses and develops their existing knowledge and is flexible enough for their expansion and integration of spaces that consider their tradition, activity, local material, and their skill.

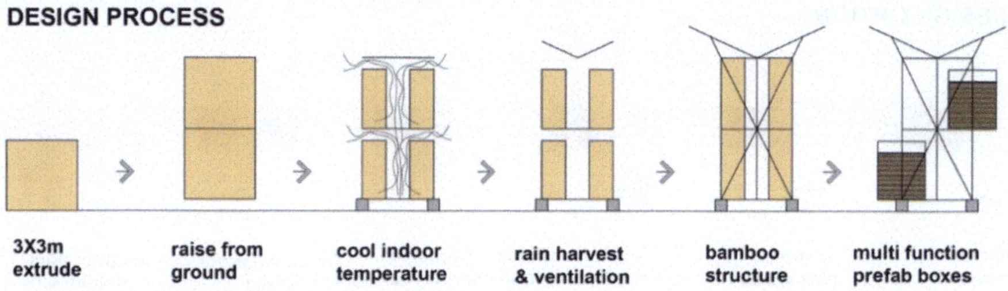

Fig. 1 Design process of in-Box House

We were looking for a design with a lot of flexibility. The house should allow for the freedom to adjust it to any need and weather type at any moment. Selecting bamboo as an appropriate building material for ease of construction, suitable for modular construction, and strong for the structural purpose would be a perfect choice for our design solution.

1.5 Design Process

Firstly, we want to use the allotted 3 m × 3 m area of the plot and then extrude the whole area up to the top of the second level which is 5 m. Since the material we have chosen for the construction of the design has been bamboo, we must raise the whole structure from the ground in order to protect the material from weathering and insect attacks. Additionally, the house has a concrete pad footing base that makes it suitable for construction on all kinds of soil.

Areas like Ethiopia have an arid climate and have sunny weather up to eight months throughout the year. So the house should consider this passive interior cooling system. By dividing the big extruded masses into smaller sections of boxes, we try to maintain cool interior temperature by letting the warm air out and letting in cool air inside. In order to use the rainwater resource, the roof of the house is a V-shape with a deep slope that helps to harvest rainwater easily and helps to escape the interior hot air.

These design processes (Fig. 1) are then structured with a vertical bamboo structure braced with diagonal bamboo pole bracing with both side. Finally, the two prefabricated bamboo lumber boxes are inserted inside the structure in flexible and multifunctional layouts.

1.6 Design Option

In-Box House is designed with flexible and multifunctional design options starting from prefabricated portable homes to floodproof layouts (Fig. 2). This makes the

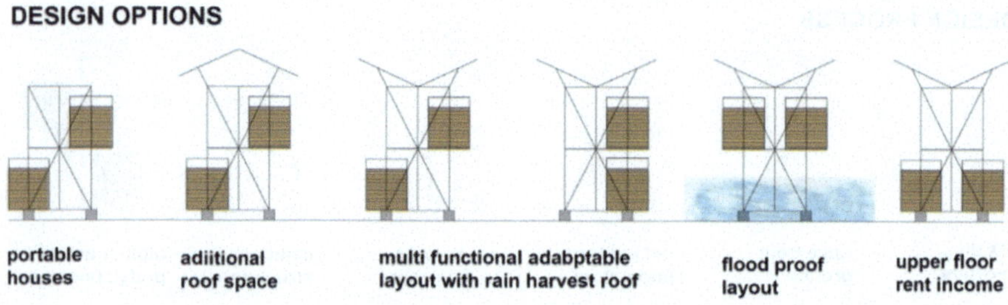

Fig. 2 Design options for in-Box House

design a viable option for the residents of the inner-city of the city that have a need for expanding and changing the plan layout according to their needs.

The boxes can be used as prefabricated portable homes as a type of mobile architecture that can be easily transported from one area to another. The demand for portable homes has been rising to solve problems like the high cost of housing in an urban area. The open area without the boxes inserted can be a rentable space in order to generate an income.

The benefit of portable boxes is that they can be assembled off-site, loaded up on a truck, and sent to another place. Since less construction work is done on-site, it can help to save money and time. The low-energy boxes will be ready in a short period of time and they can expand with additional modules. The rectangular shape of the boxes is perfect for prefabs as it fits on a truck well. in-Box House is a modern minimalist prefab that can be put on the transport or disassembled and reassembled on-site. One other interesting feature of this design is that the roof can house a home garden and has storage space for grains and other things under the roof structure.

For a country with a shortage of water resources, rainwater harvesting is a viable technology in an urban informal settlement. To take advantage of this resource, the necessary work to do is to capture the rainwater on the roof and direct it to the storage tank. The house is collecting the run-off from the roof structure surface in order to store it for later use.

With floods washing up in several parts of Ethiopia in the winter, there is a lot of discussion about how to prevent and manage floods. A floodproof house design like an in-Box House must be elevated above the base flood elevation allowing water to pass through below the house to prevent hydrostatic pressure on the home.

2 Design Development

A house in Addis Ababa is not only for a living space for a certain family, rather it is a combination of living and production spaces, even for commercial activities. For the residents, a home is a place where they not only live, eat and sleep, but also find a way to generate an income by using the space for production and business. Because of this unique way of living, there is a huge practice and demand for the extension and integration of spaces.

They often add an additional space from the upper, horizontal, and bottom sides of the existing structure of the house. This kind of addition and integration of spaces is an existing practice that helps them for easily repurposing the house according to their changing demand (Fig. 3).

We use this kind of addition and integration of spaces from existing practice and make the design more adaptable, flexible and scalable enough to meet their changing demand. By creating a simple main frame structure, and two individual boxes that can be arranged separately and in different layouts for different purposes, we solve the flexibility demand of addition and integration of spaces.

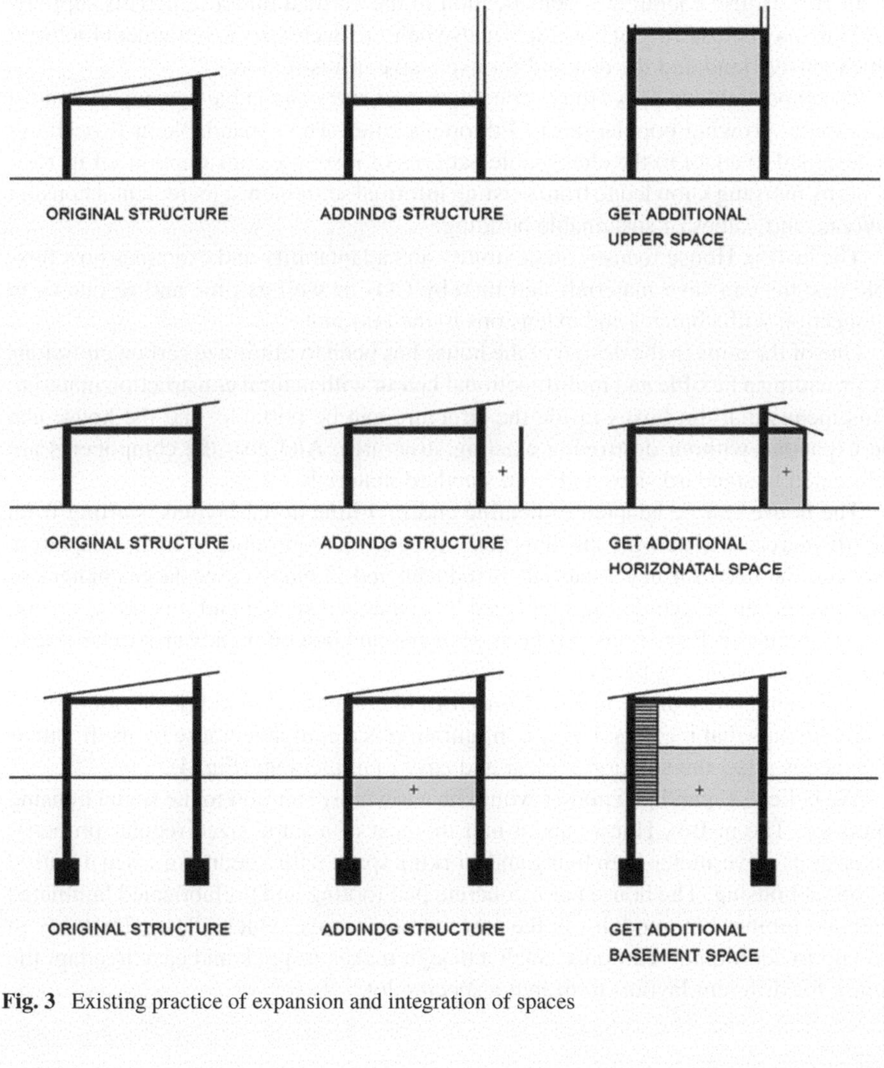

Fig. 3 Existing practice of expansion and integration of spaces

In-Box House is a multifunctional bamboo garden house designed to be constructed in a combination of bamboo poles with prefabricated laminated bamboo lumber. The prefab modular boxes are then inserted into the bamboo structural frames in many multifunctional layouts according to the functional purposes. The idea is that the house can be easily adjusted to any weather type, mood, or occasion. When the weather is sunnier, the box can be rotated against the sun direction to enjoy maximum light and experience nature to the fullest, still protected from rain and wind by the overhanging roof cover.

The house is characterized by these three dominant features expandability, scalability, and flexibility. The house is designed to function for different purposes like for garden houses, social housing, and pop-up rental shops for economic development for low-income communities.

In-Box House encourages densification in the vertical dimension. This supports the benefits of clustering of housings. It also helps to decrease the settlement footprint on expensive land and the demand for expensive infrastructures.

The expandable in-Box House offers low-cost and sustainable housing options to the rapidly growing populations of Ethiopia's cities. The expandable in-Box House is designed to adapt to the changeable patterns of resource consumption, of its residents by merging knowledge from existing informal settlements, incremental housing layouts, and values of sustainable building.

The in-Box House focuses on flexibility and adaptability and examines how flexible designs can save materials and thereby CO_2 as well as time and resources in connection with changes and extensions to the building.

One of the aims in the design of the house has been to minimize carbon emissions by providing a flexible and multifunctional layout with natural construction material. This means that the boxes inside the structure can be portable, and the house can be expanded without destroying existing structures. And also the components are fabricated in standard sizes and from standard materials.

The house can be adapted to flexible and multifunctional layouts, starting from family increases to new income thus generating work requirements. In the adaptation process, the need for new materials is reduced and in many cases, large changes or expansions can be avoided and replaced by a changed spatial and functional layout. The adaptable in-Box House has been designed and laid out in advance to be able to handle these changing needs as they arise.

The adaptability of the in-Box House lies in the many possibilities for new functional layouts that it offers. Every configuration is a complete house by itself, but as new needs arise, the solution is clear and easy to implement (Fig. 4).

We believe expandable homes would be a low-cost solution to the social housing shortage. The in-Box House, about half the cost of a same-sized regular property, consists of a five-meter-high living and working space with a bedroom when it is used as social housing. The house has a concrete pad footing and prefabricated laminated bamboo lumber box modules in the surrounding rooms, which allows flexibility to add up to additional bedrooms. Such a design makes it quick and easy to adapt the house for different layouts to fit into a specific lot.

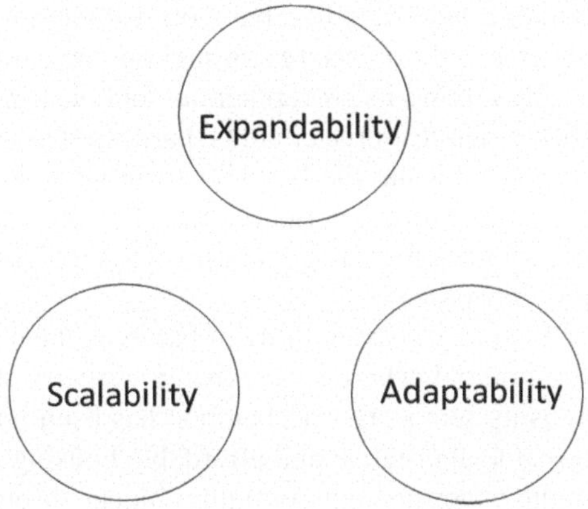

Fig. 4 Three characters of the proposed design

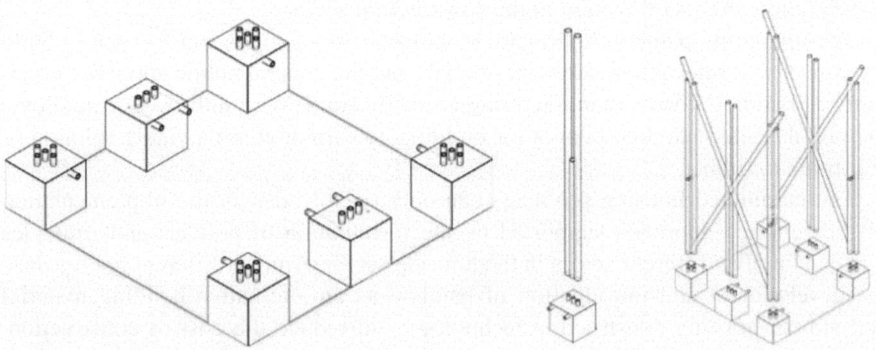

Fig. 5 Structural illustration

As Fig. 5 shown, the house uses a frame structural system with diagonal bracing support. The loads from the boxes are transferred to the columns through horizontal supports and then to the pad footings. The diagonal supports further strengthen the frame structure. The flexible connections and the material property of bamboo make this structure resilient to earthquakes.

3 Construction

3.1 Bamboo as an Alternative Material

The need for sustainable building materials is high in Ethiopia. Timber is scarce and the other building materials such as concrete are relatively expensive and may not be environmentally friendly enough. This indicates a greater potential for alternative building materials like bamboo. Ethiopia has a great amount of bamboo covering about 1 million hectares throughout the country [1].

This shows its great potential for housing and the construction sector in general. If such a rich resource could be harnessed as an alternative building material, it would greatly alleviate the building material scarcity problem in the country. Therefore, knowledge and experiences with bamboo building technology from other countries like China and Asian countries will be highly useful and relevant in Ethiopia in the present context.

Despite the availability of bamboo, the use of bamboo resources in the country is sub-optimal. There is no existing written policy on the application of bamboo in the housing sector. Hence, bamboo has been ignored completely by policymakers as a possible alternative and affordable housing material. Bamboo has not been utilized in the upcoming condominium blocks throughout the country. Most regional states with bamboo resources are at transitional stages of developing policies on bamboo conservation and its utilization in the construction sector.

The upcoming bamboo floorboard manufacturing factories can be used to build the capacity of relevant government officials and the general public about the possibility of bamboo boards manufacturing and utilization. With improved technology, the prefabricated bamboo boards for building construction are a viable solution or substitute to timber.

The cumulated housing problems can only be solved with the implementation of a diversified approach supported by the formulation of policies and strategies that can involve different actors in the housing sector. Among different approaches, the development and introduction of bamboo as an alternative building material and simple housing construction technology can reduce the cost of construction. Bamboo will also remain an important resource in the construction of recreational high-standard facilities.

3.2 Construction Process and Details

This is a prefabricated handcrafted constructive system using bamboo poles and prefabricated boxes made with laminated bamboo lumber as the main elements, to be built on a 3 m × 3 m area, with conditions similar to those in China, or housing problem commonly seen in Ethiopia.

The self-supporting prefabricated boxes made with laminated bamboo lumber are of utmost importance in this constructive system, using a laminated bamboo lumber structure (2 m × 2 m) finished with weatherproof paint on all sides. A laminated bamboo lumber panel can be made in a day; thus, in few days, between thirty and forty panels can be made that can be assembled to make the boxes. The floor, the inside roof ceiling cover, and doors are made from laminated bamboo lumbers. The roof covering is made with corrugated metal sheets. Four non-trained persons can build the house on-site in three days.

However, there are some disadvantages which should be pointed out: in our design, we planned to use laminated bamboo lumber to make the boxes. However, a new material with higher density, bamboo scrimber was used to construct the real structure. Bamboo scrimber is a little heavy after assembled to make the boxes, as this would increase the difficulty to lift it upwards on the construction of the house, making it hard for the workers. The house must be improved, by making the engineered bamboo material more lightweight and easy to assemble, disassemble and reassemble on-site where the typical beneficiaries of such houses are located.

In spite of these disadvantages, the production capacity, the lower cost, and the small construction time make it the viable alternative at this time for people living in conditions of extreme poverty.

As shown in Fig. 6, the house is to be constructed in three days with four people.

Day 1. For the first day, the primary thing to do will be planning and preparing the site for construction by clearing and removing dirt materials from the site and make the landscape slope appropriate for the building footing layout. Then collecting and bringing building materials, such as bamboo and precast concrete pad footings, and construction tools that are simple to use and can operate by hand without the use of electricity. Simple tools make the construction process easy and self-constructible without skilled labor. The next step will be constructing the two bamboo frame

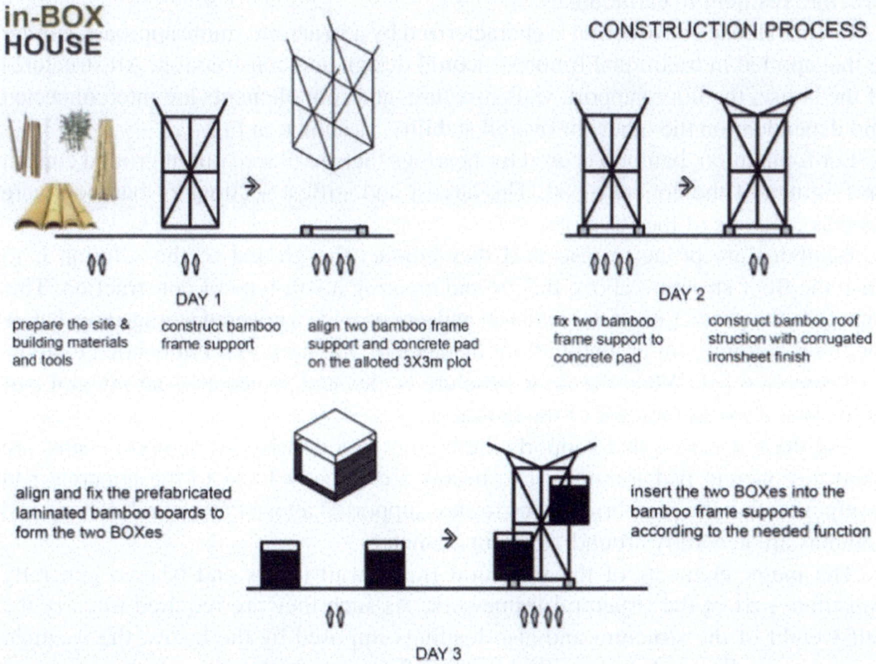

Fig. 6 Construction process of in-Box House

support on the sides of the building and align the two bamboo frame support with the concrete pads at the ground on the allotted 3 m × 3 m plot.

Day 2. On the second day, the work is to insert and fix the two bamboo frame supports into the concrete pad footings. Then construct the bamboo roof structure and fix the purlins. Finally, finish the roof with a corrugated iron sheet by fixing it with j-bolt nails. We build the structure from scratch by using raw bamboo poles with low tech construction tools not requiring high construction skill. Finally the two boxes made from prefabricated laminated bamboo lumber were transported from the company where they were fabricated to the site.

Day 3. On the last construction day, the final work to do will be to align and fix the prefabricated laminated bamboo boards to form the two boxes that are going to be inserted in the frames later. Then transport the prefabricated boxes from the factory to the site location and then insert the two boxes into the bamboo frame supports according to the needed function. The modular boxes can be laid out in a variety of ways in order to best fit the users' needs whether that be extra living space, grain storage, or floodproofing.

The house uses a frame structural system with diagonal bracing support. The loads from the boxes are transferred to the columns through horizontal supports and then to the pad footings. The diagonal bracing supports further strengthen the frame structure. The flexible connection and the material property of bamboo make this structure resilient to earthquakes.

In-Box House construction is characterized by a structural frame approach similar to that applied in traditional frame structural design and construction. All structures of the house, the floor support, wall structure, and roof elements are interconnected and dependent on the other for overall stability, as shown in Fig. 7.

For foundation, bamboo is used for bearings that are placed out of ground contact on footings of the concrete pad. The largest and stiffest sections of bamboo were used as columns of the structure.

Bamboos are prone to insects if they contact the ground so the solution is to raise the floor structure above the ground creating a stilt type of construction. This improves the protection of the material and can provide a covered storage area below the floor. A minimum ground to floor distance of 500 mm is recommended to allow for inspection [3]. When the floor structure is elevated, it becomes an integral part of the structural framework of the building.

The floor structure that supports the bottom box consists of bamboo beams are fixed to concrete pad footings. The beams are directly fixed to the concrete pad footings so that the prefabricated boxes are supported above the ground. Beams and columns are generally around 10 cm in diameter.

The major elements of the structural frame wall (posts and beams) generally constitute part of the structural framework. As such they are required to carry the self-weight of the structure and also loadings imposed by the boxes, the weather, and occasionally, earthquakes. To this end, efficient and adequate jointing was used when constructed.

The vertical frame structure of the house consists of vertical space bamboo pole columns braced diagonal in both directions and as well on both sides. At the second

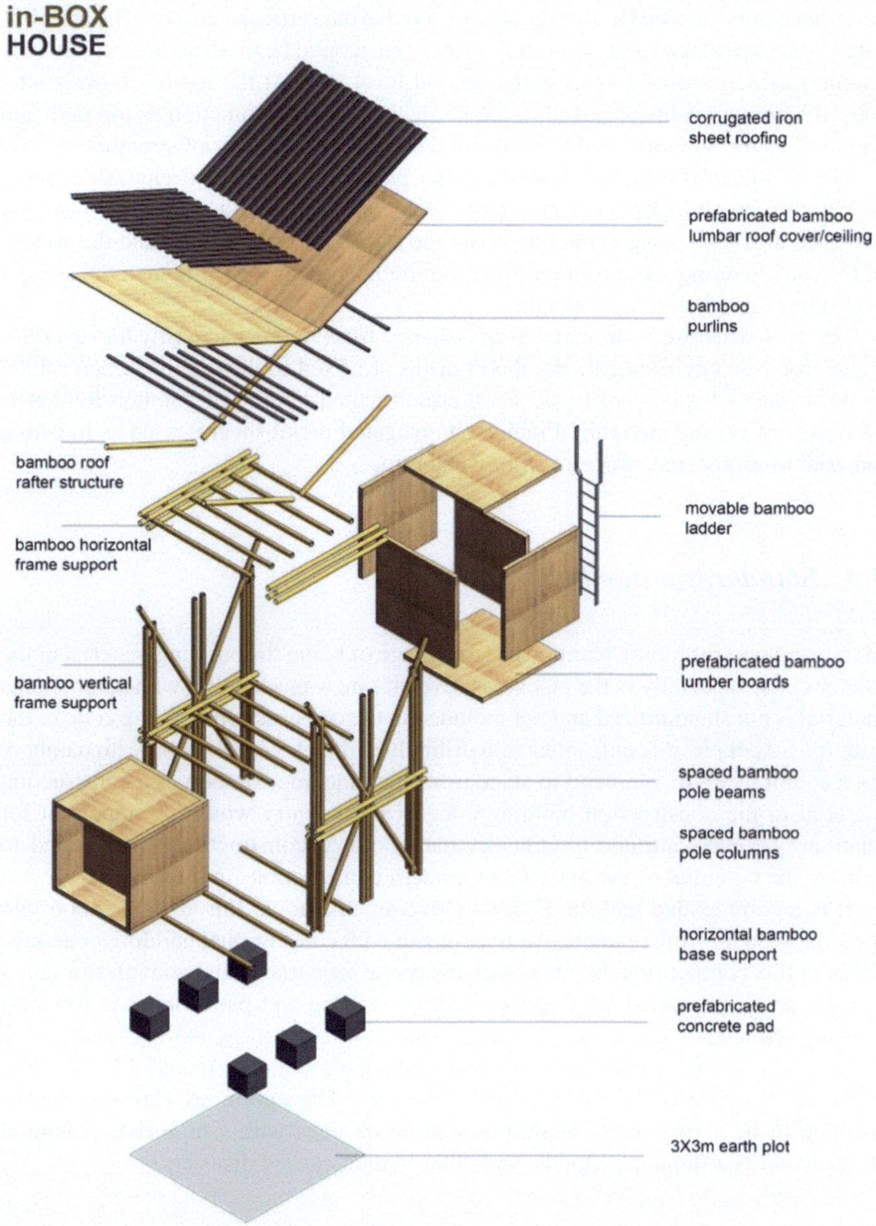

Fig. 7 Exploded illustration of in-Box House

level, horizontal spaced bamboo beams are fixed to the vertical columns. The prefabricated boxes are placed and supported by these horizontal beam structures. A movable bamboo ladder is used to access the second-level box. At the top level, horizontal spaced bamboo beams are fixed to the vertical columns to function as top tie beam and used as the support of additional storage space below the roof structure.

The roof of the house is designed to offer protection against extremes of weather including rain, sun, and wind, and have usable space below its covering. The roof structure must be strong enough to resist the forces caused by wind and the weight of the roof coverings. For roof purpose, bamboo is best as a roofing material since it is strong, resilient and light-weight.

The roof structure is designed as a V-shaped truss in order to easily harvest rainwater from the environment. Bamboo purlins are fixed to the main bamboo rafters to form the V-shaped roof truss. Then prefabricated laminated bamboo lumber is used as roof ceiling covering. Finally, a corrugated metal sheet is used as finishing material to protect the structure from weathering.

3.3 *Standardize Bamboo*

In our country (Ethiopia), bamboo has the image of being the building material of the poorer class, especially in the city areas. Even if one wants to build with bamboo, the material is not standardized and not included in the construction building code of the country. So people are confronted with difficulties if they want to build with bamboo. As a solution, we recommend to standardize and include bamboo as a construction material in the construction building code of the country would be important for changing people's attitude to consider bamboo as a construction material and to explore the potential of the material in modern constructions in the city.

It is recommended that the Federal Government should introduce bamboo into the Ethiopian housing market so as to compete with conventional building materials. Without this competition, bamboo will always be seen traditional and inferior material. It can be promoted for flooring, ceiling, walling and partitioning in low-cost housing schemes.

Bamboo will be a more low-cost and sustainable building material for the city especially to curb the housing issue of the city. The enormous elasticity makes bamboo to be a very useful building material in areas with a high risk of natural disasters like earthquakes, floods, and other human-caused disasters.

4 Experience

The 2019 International Bamboo Construction Competition (IBCC2019) brought us our first-ever chance to travel abroad, carrying an exhilarating assignment of building and showcasing our bamboo house in the INBAR Garden at the International Horticultural Exhibition 2019. Right from the moment we arrived, the hospitability and warm reception from the entire INBAR team was something we will not pretermit.

What sets out "in-Box House" from the rest finalists was the fact that it was to be built from raw bamboo, with low tech construction skills, and from scratch within three days. In those three precious days we have learned, exchanged knowledge on construction skills and worked hand in hand with the local Chinese bamboo craftsman which we would like to give a big massive thanks to and most importantly finished building "in-Box House" which was a huge success in the parameter of practicality, flexibility, and precision of our prototype. It also served as a case study for future projects that can be attained with a short timeline, differentiating climate and flexibility.

On the third and final construction day of the competition, we encounter difficulty in putting the upper prefabricated bamboo box at the top structure of the house; it was heavier than we initially thought it would have been. We try some techniques to lift the box to the upper level of the structure but with our construction technique and tools, it was too heavy to manage. At last, we manage to adapt the design situation by repurposing the function of the box. The design by itself allows for more flexibility to adapt to any situation and repurpose the designed spaces according to the context. From this experience, we learn that you can put and arrange the individual boxes near the main structure and use them for another purpose by claiming extra spaces. One can still see the strong relationship between the main structure (the host) and the box; they belong to each other even if they were separated from the original location. In the future, we are thinking of making the boxes lighter for the ease of moving and managing them.

IBCC 2019 brought us a great opportunity to share ideas, brainstorm, and learn from the rest of the finalist teams, but also gave us a chance to experience a new culture and lifestyle. Our stay in Beijing and at the International Horticultural Exhibition 2019 was an eye-opener that helped us know more about different cultures, lifestyles, building systems and architecture. We were astonished by the overall hospitality of the whole INBAR team and the people in Beijing. Having this opportunity, we would like to thank everyone who shared this incredible journey with us.

5 Renderings and Photos

Several renderings of the design (Fig. 8) and photos of the real building (Fig. 9) are provided as follows.

Fig. 8 Renderings of in-Box House

Fig. 9 Photos of in-Box House

Acknowledgements We would like to acknowledge all financial and technical support of the International Bamboo and Rattan Organization (INBAR) and the staff for giving us the chance to build our design proposal at the International Horticulture Expo 2019 in Beijing by covering all the expenses. We would like to extend our deepest and sincerest gratitude to all the people who helped there in any manner, who have shared the effort and knowledge in order to make this project a reality. We are over helmed in all humbleness and gratefulness to acknowledge our depth to all those inside INBAR and the Organizing Committee of IBCC 2019, who have helped us to put these projects, well above the level of simplicity and into something concrete. We were astonished by the overall hospitality of the whole INBAR and IBCC 2019 team and the people in Beijing. Having this opportunity, we would like to thank everyone who shared this incredible journey with us.

References

1. Jacob, K.K., and E.M. Sylvia. 2011. The value chain development and sustainability of bamboo housing in Ethiopia. *INBAR*. 10: 15.
2. UN Habitat. 2007. Situation analysis of informal settlements in Addis Ababa. UN Habitat. 22.
3. Janssen, J.J.A. 1995. *Building with bamboo, a handbook*, 2nd ed. London, UK: Intermediate Technology Publications. Southampton Row.

Machine of Bamboo

Huimin Wu, Sicheng Zhou, and Xiaonan Li

Abstract The design philosophy is adopted as "Machine of Bamboo". The systematic design and construction of this work are based on engineered bamboo, connected with metal joints. This approach can make the prefabrication of elements in the factory and installation of them on-site come true. This up-to-date design concept aims to develop and apply new materials and new technologies into green buildings.

Keywords Engineered bamboo · Lattice column · Construction · Prefabricated

1 Concept Design

The design philosophy is adopted as "Machine of Bamboo". Engineered bamboo is adopted as one of the construction materials. Considering that the mechanical properties of raw bamboo culms, known as anisotropy, is greatly affected by its moisture content, therefore the durability of which is poor. To tackle these problems, engineered bamboo has been invented [1], which could be with mass production and application. Furthermore, mechanical performances of engineered bamboo under static and dynamic loading conditions have been studied by Xiao et al. [2], Zhou et al. [3, 4], and Wang et al. [5], showing that bamboo has great potential as a building material. A house is a machine for living in Corbusier [6]. "Machine" means prefabricated, reproducible and fast propagation. Bamboo material has the ability to meet the basic durability of the buildings, embodying the nature of the machine. It is a future-oriented ecological building material, which is more suitable for future sustainable development in the background of social industrial production.

The pre-specified dimension is based on a foundation of 3 m × 3 m. The works, the pavilion building, built on the foundation is 5 m height, with projected area of 3.75 m

H. Wu · X. Li
School of Architecture, Nanjing University, Nanjing, China

S. Zhou (✉)
Department of Architecture, Built Environment and Construction Engineering, Politecnico di Milano, Milan, Italy
e-mail: sicheng.zhou@polimi.it

K. Liu et al. (eds.), *2019 International Bamboo Construction Competition*, Springer Tracts in Civil Engineering, https://doi.org/10.1007/978-3-030-91990-0_7

× 3.75 m. The building is made of engineered bamboo. According to its performance, industrial process, and module of engineered bamboo, 58 mm × 58 mm and 38 mm × 38 mm was adopted as sections of the main and the secondary bearing columns respectively. And 120 mm × 38 mm and 80 mm × 38 mm was adopted as sections of the main and of the secondary beams respectively. The cross-section size of 38 mm × 38 mm was selected as the section of auxiliary transverse members, such as the elevation fences. Based on the principle of the convenience of material processing and the maximization of material properties, the section size is finally determined through repeated tests, which fully embodies and makes the most of the longitudinal mechanical properties of industrial bamboo.

The main structure of the building consists of 3 frames, composed of 10 main columns. Two of the three frames are the main bearing parts, which are marked in red in Fig. 1a. And Fig. 1 shows the design logic, with different colors manifesting

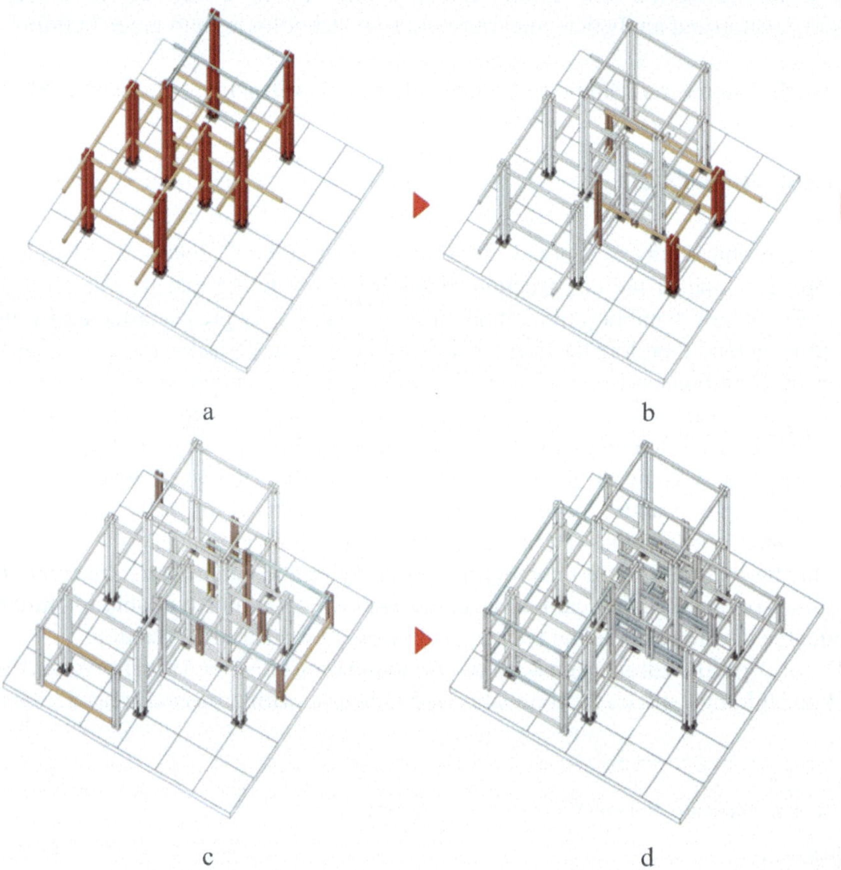

Fig. 1 Structure logic. Main structure (**a**); Accessory structure (**b**); Cantilever structure (**c**); Auxiliary component (**d**)

Fig. 2 Model of main architecture structure

different components: main columns (red), secondary columns (orange), main beams (light yellow), and secondary beams (light blue). The design concept is that the structure regards two main frames, which have different top elevations, as the core. The third frame with a cantilever is implanted. Three frames of different heights make the whole structure rich in layering. The secondary columns, secondary beams, stair stringers, and ledges are used to set up auxiliary parts and improve the whole performance of the structure.

After the structure is designed, a manual model is made to simulate its loading conditions (shown in Fig. 2). The simulation shows that the connection of the two main frames is weak and one additional main beam (with the section of 120 mm × 38 mm) was attached to increase its stability.

2 Design Development

Considering that the base area is 3 m × 3 m, the layout scheme adopts 0.75 m as the basic module. The projected area of the structure is 3.75 m × 3.75 m. According to the limitation of the height (5 m), human accessible height is adopted as 3.6 m, with a local cantilever of 0.75 m (Fig. 3). The building consists of two main parts and one accessory part. The two main structural parts are divided into platform units and aisle units, connected with the accessory part. The platform units include two sightseeing zones (one inside at the elevation of 1.2 m (Fig. 4) and the other outside one at the elevation of 3.6 m (Fig. 5). A seating area is also set up at the elevation of 0.48 m of the accessory part.

The building plane is developed with 0.75 m × 0.75 m as the module and is composed of three major 1.5 m × 1.5 m units. The building facade is expanded with a modulus of 0.75 m × 1.2 m, with a human-accessible height of 2.4 m.

The facade is made of sunlight panels, as shown in Figs. 6 and 7, which are easy to install and ensure privacy and good architectural lighting. Two kinds of translucent and transparent sunlight plates were used in the main structure and the accessory corridor part respectively, strengthening the block sense of the whole building. Prefabricated steel-trellis plates were adopted as the floors and treads.

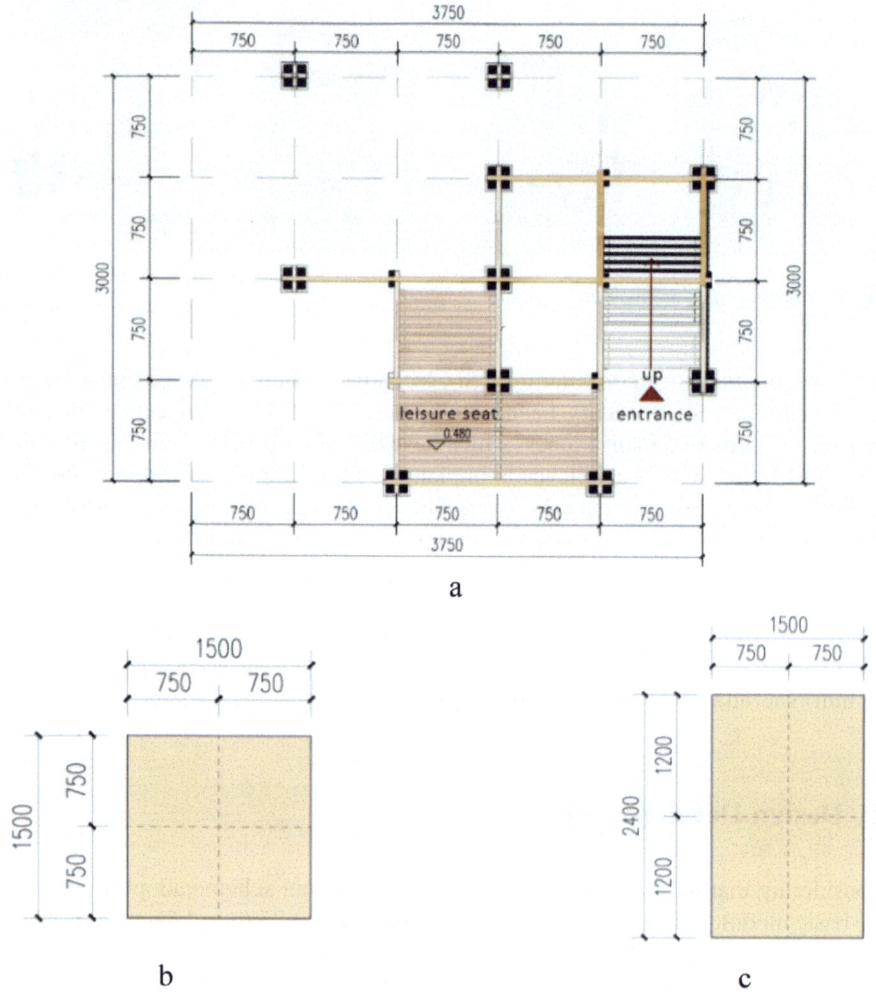

Fig. 3 Plane of 0.75 m in height (**a**); plan modulus (**b**); section modulus (**c**)

3 Construction

3.1 Structural Details

Bolts are used as connecting members between industrial bamboo poles, and prefabricated metal parts are used to reinforce the main stress-concentration joints. There are two types of bolts used according to the design scheme: M8 and M10. In addition, M12 chemical anchor bolts are used to connect the columns with the base. According to the protection distance requirement, the hole diameter of M8 and M10 bolts on the

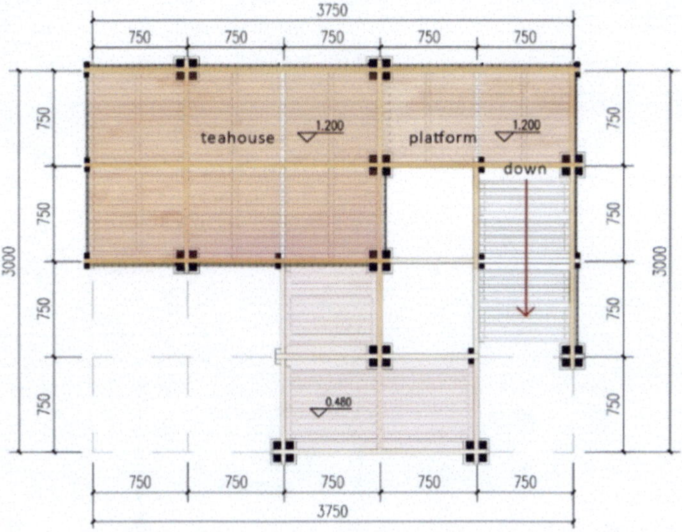

Fig. 4 Plane of 1.2 m in height

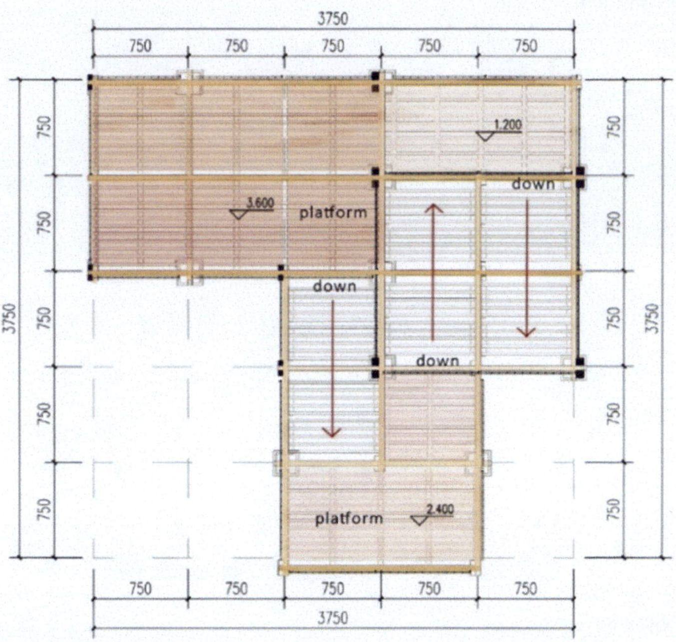

Fig. 5 Plane of 3.6 m in height

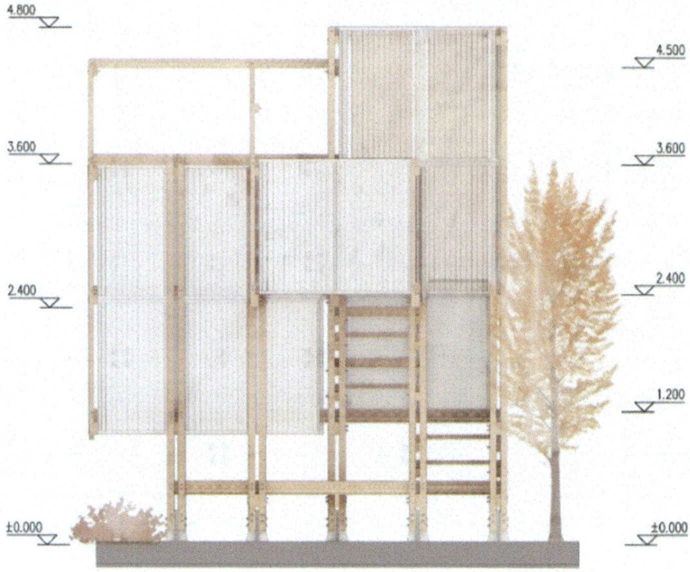

Fig. 6 Front elevation

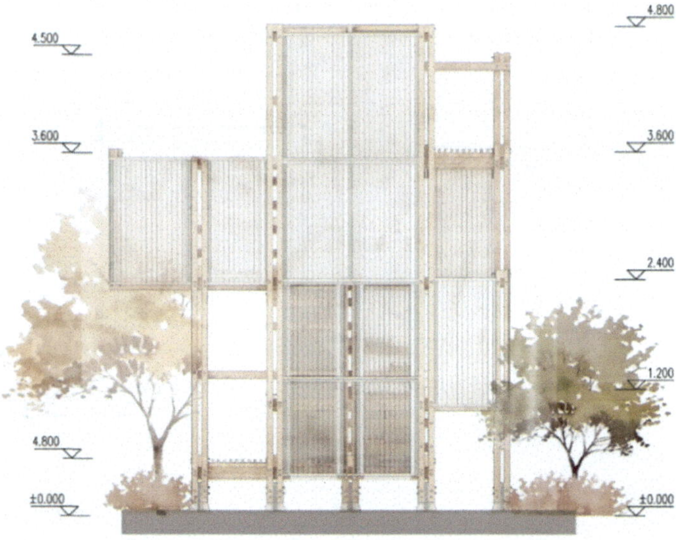

Fig. 7 Side elevation

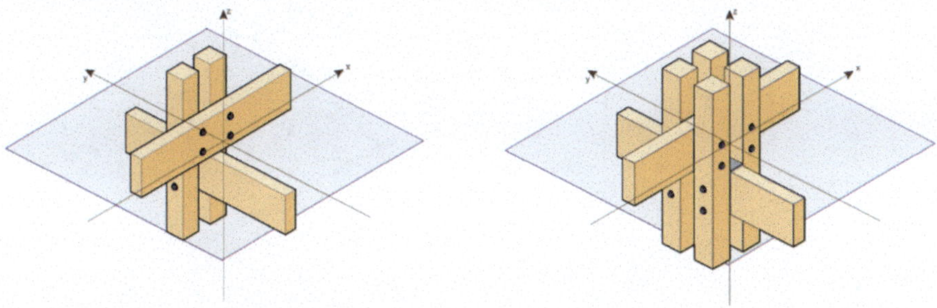

Fig. 8 Details of engineered bamboo joints

rod are 9 and 12 mm, and the distance between the center points of the bolt holes is 60 mm, from the rod boundary 30 mm. Thus, the main beam with a section height of 120 mm can be connected with the vertical bar through 2 bolts, while the secondary beam with a section height of 80 mm can be connected with the vertical bar through 1 bolt. The details are shown in Fig. 8.

The steel-trellis plates are used as the building floor, and the structural connection is carried out by conventional industrial practice. The building facade is made of simple and economical sunlight plates as the enveloping, which is fixed with bamboo poles on the side by 2 mm thick angle aluminum and self-tapping screws, shown in Fig. 9.

The structure was modeled, adopting information reported in Tables 1 and 2. Then, the structural system was analyzed by using SAP 2000 and the results are shown in Fig. 10.

Through the calculation, the stress and deformation limitation can be controlled under allowable value, and the structure meets the requirement of safety and serviceability.

3.2 Construction Procedure

The construction process is divided into two parts: pre-construction (Fig. 11) and on-site construction. The pre-construction was conducted in the manufacturing factory, to check the feasibility of the design scheme. It was done by workers in the factory, assembling the main structure. And the lattice columns were assembled and packed before transporting to the INBAR garden at the 2019 Beijing Expo.

The on-site construction was carried out by six students and two workers, taking three days. On the first day, the chemical bolts were anchored in the isolated footing. After the curing of the chemical glue, the main columns were set up and the perpendicularity was calibrated by the laser gradient. The connection component, angle aluminum, was installed, which was used to connect steel-trellis plates with main beams and connect sunlight panels with the main columns. Most of the beams and

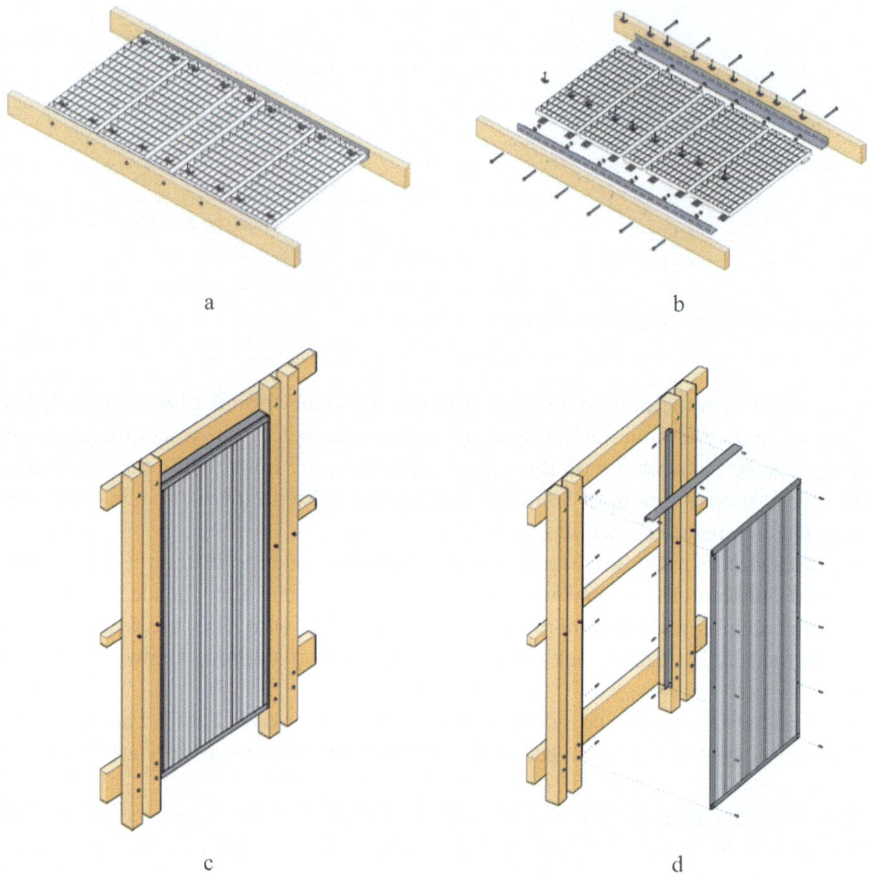

Fig. 9 Steel-trellis plates (**a**); exploded view of steel-trellis plates (**b**); PC sunlight panel (**c**); exploded view of PC sunlight panel (**d**)

secondary columns were assembled on the second day. During the last day, other components, including the steel-trellis plates and sunlight panels were installed. The procedure is shown in Fig. 12.

Day 1 of the construction: There are many problems in the construction of the foundation part: the uneven ground, the rough foundation construction, and the rain make it more difficult to put the chemical anchor and bolt reagent into the foundation, which directly leads to difficulties in the construction of the foundation on site. The chemical anchor and bolt could not be installed until the rain stopped.

The installation process (Fig. 13) includes four steps: (1) Layout the lines on a separate basis to determine the positions of the four chemical anchors on each base and to ensure that their X and Y directions are on the same line; (2) After curing at each anchor point, chemical anchor bolts were installed; (3) The top surface of the base was calibrated: double nuts are used to calibrate the plane and the level is used

Table. 1 Load analysis

Design load		
Load name	Load value (kN/m^2)	Label
Dead load	1.0	D
Floor live load	1.0	L_f
Roof live load	1.0	L_r
Stair live load	1.0	Ls
Load combination		
Ultimate limit state	$1.35D + 1.4 \times 0.7 \times 1.0L_f$	
	$1.35D + 1.4 \times 0.7 \times 1.0Lr$	
	$1.35D + 1.4 \times 0.7 \times 1.0Ls$	
	$1.35D + 1.4 \times 0.7 \times (1.0L_f + 1.0Lr + 1.0Ls)$	
	$1.2D + 1.4 \times 1.0L_f$	
	$1.2D + 1.4 \times 1.0L_r$	
	$1.2D + 1.4 \times 1.0Ls$	
	$1.2D + 1.4 \times (L_f + Lr + Ls)$	
Serviceable limit state	$D + L_f + Lr + Ls$	

Table. 2 List of elements

Section information	Unit: mm
Label	Size
Column 1	120 × 120
Column 2	60 × 120
Column 3	40 × 80
Beam 1	40 × 80
Beam 2	40 × 120

for measurement; (4) After the chemical anchor solidified, the main columns were installed.

Day 2 of the construction: According to the number of members, the main beams were installed. The frame of the whole building enhances the structural stability in the construction process. The bolts were not tightened, to adjust the construction error.

Day 3 of the construction: The main work of the last day was to install all the steel-trellis plates (Fig. 14a, b) and sunlight panels (Fig. 14c).

At the entrance of the building, two steps were made with red brick and slate (Fig. 15). Meanwhile, gravel blocks in the site were used to lay a stone road into the building, and red bricks were used to draw the edge.

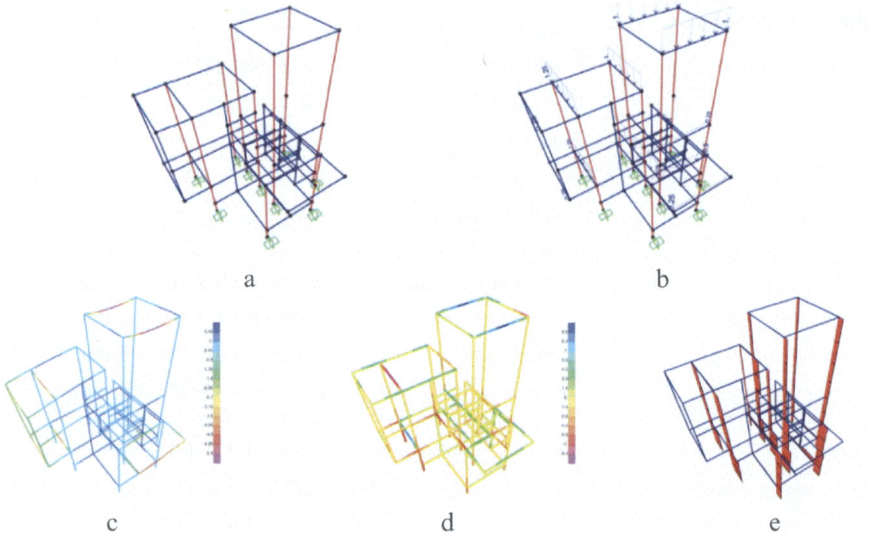

Fig. 10 Structural analysis model set-up (**a**); applying loads (**b**); deformation (**c**); stress distribution (**d**); axial load distribution (**e**)

Fig. 11 Pre-construction in the factory

Fig. 12 Different stages of assembling on-site. Day 1 (**a**); Day 2 (**b**); Day 3 (**c**)

Fig. 13 On-site installation: lay out lines (**a**); drilling holes and installing bolts (**b**); calibration with the level (**c**); adjust double nuts (**d**)

4 Experience

The experience of three-days construction provides us a lot of harvest and feelings. The difficulties in the construction stage were different from those we encountered in the design stage. There were many unknown factors during the construction, such as the difficulty of multiple forms of elements, deficiencies of materials and tools,

a b c

Fig. 14 Installing steel-trellis plates (**a**, **b**); installing PC sunlight pannels (**c**)

Fig. 15 Red brick drawing the edge at the entrance tread

complicated installation and etc. In addition, the rain also made our work tougher. However, as each member of us has a clear division of labor, everyone works very hard and earnestly so that we can overcome all the difficulties to complete the construction, which is our biggest advantage. It was a great experience for us to see our design from scratch, the rods, bolts, solar panels, and steel panels, to the final completion of an amazing building. Although there remained several points that needed to be improved, it had taken the first step of applying engineered bamboo as the main building structural material into practice, which is future-oriented. Next, the bamboo should be excavated at a higher height to build a higher level of the industrial bamboo building.

5 Renderings and Photos

Several renderings and photos of the real building are provided in Fig. 16. Figure 16a, b show the architectural rendering, which can be used as pavilions in the parks, scenic areas, etc. It is comfortable and enjoyable to have a seat and enjoy the sunshine, which shall be the top choice of a leisurely afternoon. Besides, Fig. 16c, d are the real photos of the architecture during the daylight and night. The original architecture is basically the same as the rendering.

Fig. 16 Renderings in daytime (**a**); renderings at night (**b**); photo of the real building in daytime (**c**); photo of the real building at night (**d**)

References

1. Xiao, Y., Q. Zhou, and B. Shan. 2010. Design and construction of modern bamboo bridges. *Building Structure* 15 (5): 533–541.
2. Xiao, Y., R.Z. Yang, and B. Shan. 2013. Production, environmental impact and mechanical properties of glubam. *Construction and Building Materials* 44: 765–773.
3. Zhou, S.C., C. Demartino, and Y. Xiao. 2020. High-strain rate compressive behavior of Douglas fir and glubam. *Construction and Building Materials* 258: 119466.
4. Zhou, S.C., Chu, F.Z., Lv, X.H., and Xiao, Y. 2021. Experimental studies on glubam columns under axial compression. Journal of Building Engineering, 103453.
5. Wang, J.S., C. Demartino, Y. Xiao, and Y.Y. Li. 2018. Thermal insulation performance of bamboo-and wood-based shear walls in light-frame buildings. *Energy and Buildings.* 168: 167–169.
6. Corbusier, L. 1923. *Vers Une Architecture.* Paris: Editions Cres.

Bloom

Xin Zhuo, Chidi Wang, Chenchen Yang, Guanbao Rui, and Ting Huang

Abstract Because of the low mechanical performance of the joints and the low transverse strength of bamboo culms, the traditional bamboo structure can only be used on small buildings. With an innovative topological relationship between bamboo culms, we have invented a new type of structural system named "Multi-culm Bamboo Spatial Lattice Structures". The bamboo culms are assembled end-to-end to form a triangular unit, and then adjacent bamboo culms are connected in a triangular parallel way in construction. The highly efficient method to construct the bamboo pavilion from the structural center to the edge is applied with the help of a tripod and a hand chain hoist. Technologies of this new type of structural system can be applied to any country or region, because of the simple processing of the components, the low cost and convenient purchase of the connectors, and the relatively good performance of the structure.

Keywords Hyperbolic paraboloid shape · Triangle unit · Multi-culm bamboo spatial lattice structures · Structural construction method

1 Concept Design

1.1 Development Status of Curved-Surface Bamboo Culm Structure

Many landmark-structure buildings are characterized by many fluent curved-surface shapes. However, nodes of those complicated curved surfaces are mostly not the same, and their connecting bars are from different spatial directions. Their diversity and complexity notably raise the technical difficulties and increase the cost of production and construction.

X. Zhuo (✉) · C. Wang · C. Yang · G. Rui · T. Huang
College of Civil Engineering and Architecture, Zhejiang University, Hangzhou, China
e-mail: zhuoxin@zju.edu.cn

C. Wang
e-mail: chidiwang@126.com

K. Liu et al. (eds.), *2019 International Bamboo Construction Competition*, Springer Tracts in Civil Engineering, https://doi.org/10.1007/978-3-030-91990-0_8

Recently, some new type steel nodes with remarkable performance appeared for curved-surface bamboo culm structure, relying on advanced equipment to complete sophisticated production. Additionally, there still exists the challenge of diversity, multi-direction, and expensiveness [1]. At present, the process of bending shaping of bamboo culms is unable to be done in mechanization. The quality and efficiency of processing totally depend on the bamboo workers' skill and experience, which refers to factors of diameter, curvature, and moisture content of bamboo culms, and baking temperature and duration, as well as the timing and level of force-adding while baking.

In a word, the cost of joints manufacturing and bamboo culm processing is much higher than that of bamboo culm itself. The total cost relies on those advanced equipment used in the factory and the salary for skilled workers. These technical and economic reasons have limited the development of the curved-surface bamboo culm structures.

1.2 Innovation on the System of Bamboo Structures

In order to realize standardized design, industrialized manufacture and assembly construction of large bamboo structures, and base on the philosophy of "the simpler, the better", Prof. Zhuo Xin and Prof. Dong Shilin from Zhejiang University invented a *Multi-culm Bamboo Spatial Lattice Structure* [2].

With the design concept of "breaking up the whole into parts", the whole structure consists of different bamboo triangle units Fig. 1, which can be stuffed in a normal SUV that can meet one of the requirements required by IBCC 2019, "easy transportation", as shown in Fig. 2.

The design also meets the requirement of "easy installation and dismantlement". For both long-term and temporary structures, the structure bearing capacity is guaranteed firstly. Then, for the former, triangle units are designed to connect with each other through rigid connection by bolts that can be both installed and dismantled, as shown in Option A of Fig. 3. For the latter, triangle units are connected with each other through fastener connection by own-developed double semi-ring clamps that are more convenient in installation and dismantlement, as shown in Fig. 1 and Option B of Fig. 3.

1.3 Design Philosophy

In terms of design philosophy, architectural image, structural system, and technical methods, our works BLOOM synthetically echo the theme of IBCC 2019, Blossom. The design aims to rebuild the original nature, improve the sense of bamboo pavilion, and wake natural circumstances of "unlimited landscape is assembled in a pavilion". Merging the pavilion into nature, echoing its surroundings in an open attitude, a

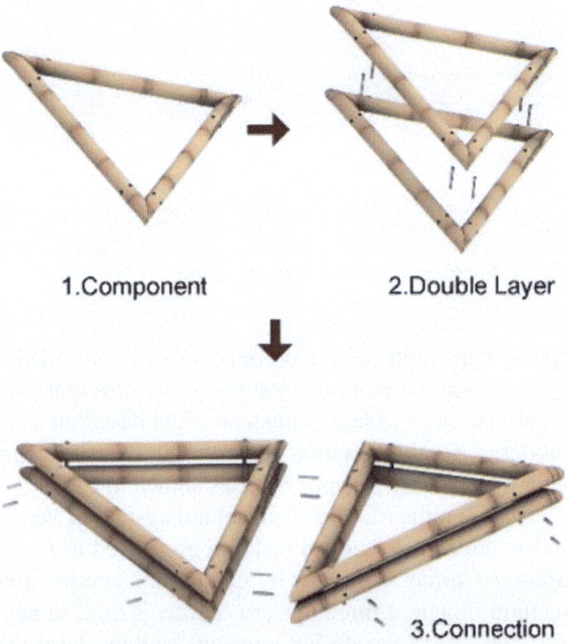

Fig. 1 Structural triangle unit

Fig. 2 Convenient carriage for structural units

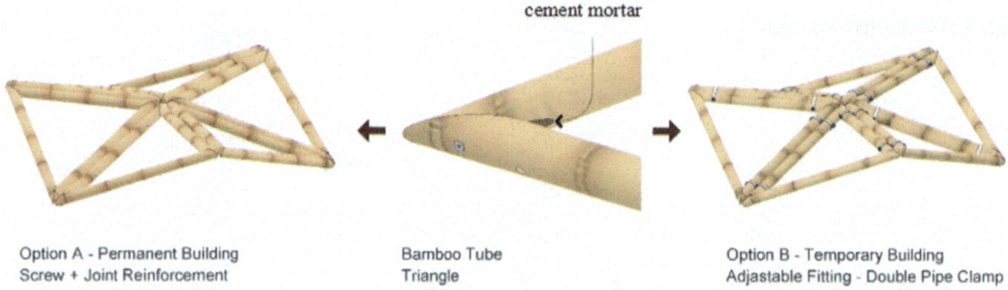

Fig. 3 Connection modes for the structure

space with multi-direction connections is provided. To respect the possibilities of natural material bamboo, and utilize its different forms, a mild style of continuous combination of triangle units is formed based on a creative structural system. In the modern city, this pavilion will rebuild a warm feeling of original shielding, under which any activity can be held, as shown in Fig. 4.

Abandon the traditional structural system of vertical posts added with covers, an independently-developed technology is used to form a system *Multi-culm Bamboo Spatial Lattice Structures* by continuous combination of triangle units. The architectural image of three upwards petals is used to respond to the competition theme *Blossom*. With the design concept "making hard things simple", the complicated curved surface of the shape of the petal is divided into 9 types of simple triangles, as shown in Fig. 5a.

Fig. 4 Renderings of BLOOM

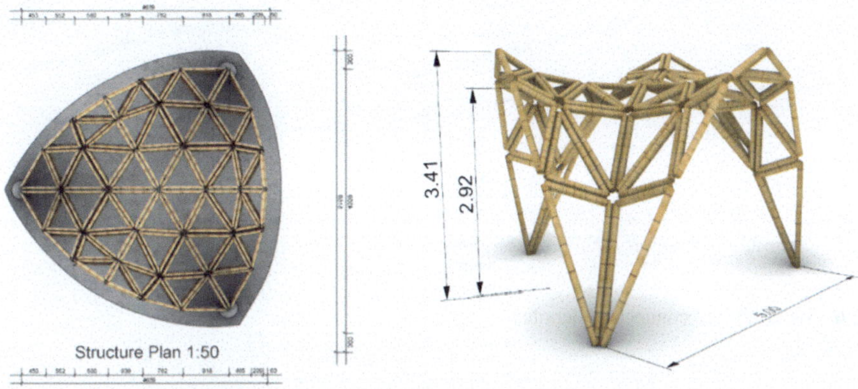

a b

Fig. 5 Structure plan (**a**); structure of BLOOM (**b**)

2 Design Development

2.1 Architectural Design

Size. The architectural plane of the pavilion is an equilateral triangle (Fig. 5a). The area of the enclosed base is 9 m^2. The surface area of the hyperbolic-paraboloid shape pavilion is 15.6 m^2 (less than 16 m^2). The height of the pavilion center is 2.92 m and the height of the eave is 3.41 m, as shown in Fig. 5b.

Materials. Bamboo culms are used as structural components, and stainless steel bolts are used as joints. Bamboo culms were treated through the patent for the invention of "lightproof anti-aging agent used for bamboo materials and its preparation method" and other mature technologies of insect prevention, mold proof, and anti-corrosion, to meet the requirement "long-term durability of structural pieces". The coiled materials were used for waterproofing. Bamboo strips were used as substructures to support the internal decorative layer and the external decorative layer of the building surfaces.

Connections. Bamboo structural components were connected by steel bolts. The internal decorative layer (bamboo strips) was fixed on bamboo culms by nails. The waterproofing layer was stick onto the internal decorative layer by an adhesive. The external decorative thatch roof was attached to the waterproofing layer by binding, as shown in Fig. 6.

Diagram of the drainage system. The shape of the structures is a hyperbolic paraboloid. The negative Gaussian curvature enables rainwater flows down to the ground via the valley line of the structure and avoid dropping directly from the eaves (Fig. 7a). The interlining of internal and external decorative surface layers uses coiled materials for waterproofing in order to meet the requirement of the competition "detailed design for keeping dry, good drainage and ventilation", as shown in Fig. 7b.

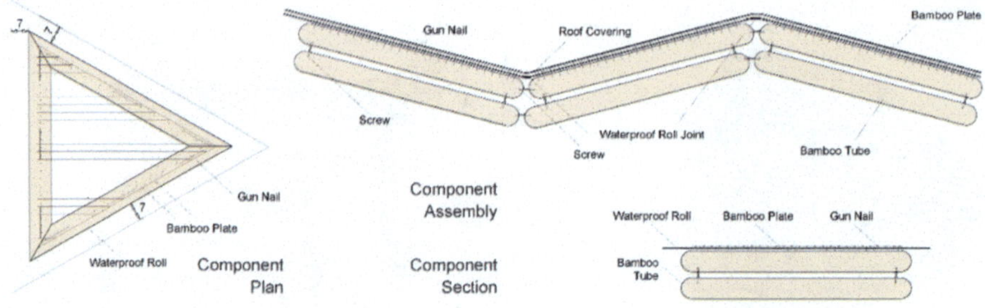

Fig. 6 Materials and connection details

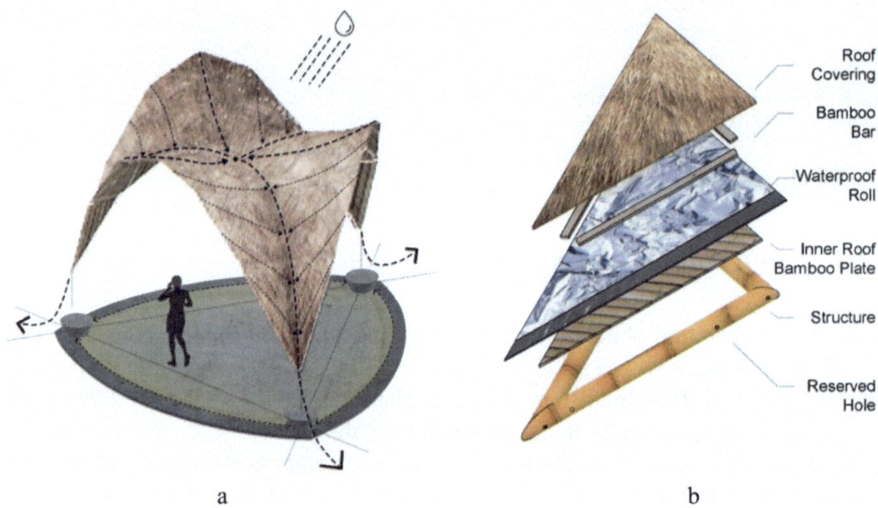

Fig. 7 Drainage system (**a**); roof details (**b**)

Structure of each layer of the roof. In order to provide shelter from sunlight and rain, as well as thermal insulation, the inner decorative layer of bamboo mats, the waterproofing layer of coiled materials, and the external decorative layer of straw were gradually assembled above the main structure, as shown in Fig. 7b.

The architectural form of the bamboo pavilion adopts a combined image of three centrosymmetric blooming petals. The open structure makes all the activities possible, allowing users to flexibly create layouts, as shown in Fig. 8.

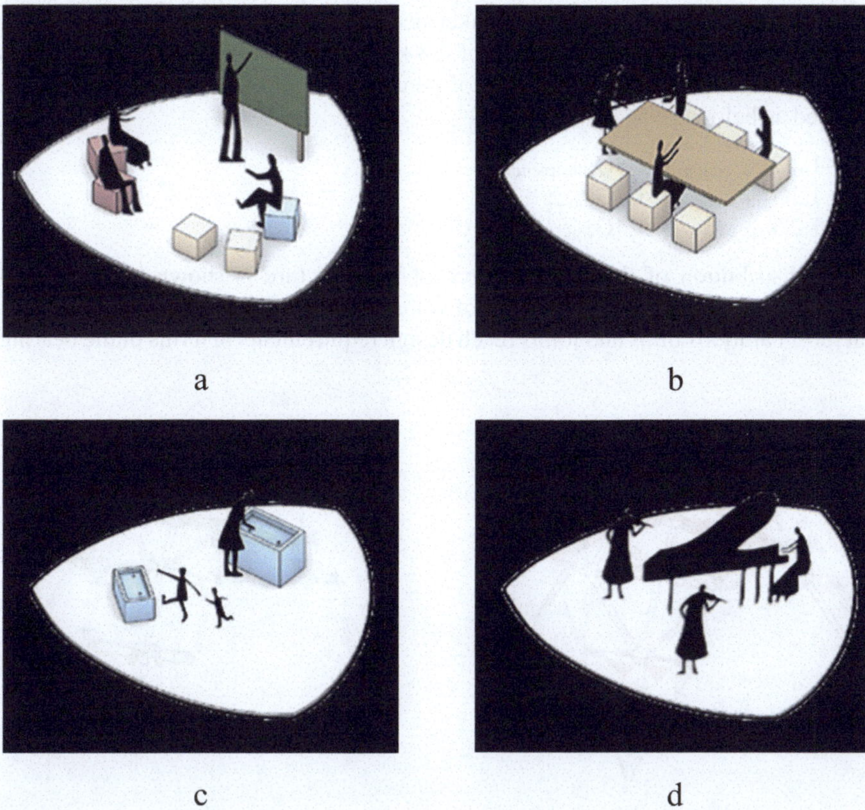

Fig. 8 Functions of the pavilion: Classroom (**a**); office (**b**); equipment (**c**); recreation (**d**)

2.2 *Structural Design*

The structure consists of 54 bamboo culm triangles units with 9 different types. The weight of each triangle unit was lighter than 10 kg which can be easily installed. The elasticity modulus of the culm is 1.5×10^4 MPa [3].

Structural components. The structures were made of straight bamboo culms of 0.5–1.9 m long with an outer diameter of 70–80 mm. No need to bend bamboo culms and we use customized multi-direction nodes, the total cost of the structure is lower that can meet the requirement of competition "no more than USD 10,000 of the total budget for all construction materials".

Joints. Stainless steel strews with a diameter of 18 mm were used as joints. Their elasticity modulus is 2.05×10^5 MPa.

Foundation. The weight of this structure is light and three footholds are less affected by uneven settlement. Therefore, three arch feet were fixed by three independent concrete foundations with a cylinder of 0.4 m deep and 0.5 m diameter.

Load. Standard values of load are taken according to the self-weight of the structure of 4 kg/m, the construction load of 4.8 kN, snow load of 0.4 kN/m^2. The sets of load of the structure in the period of construction and the period of usage are provided as below:

(1) $1.2G_{self\_weight} + 1.4Q_{construction}$;
(2) $1.2G_{self\_weight} + 1.4G_{snow}$;
(3) $1.2G_{self\_weight} + 1.4G_{wind}$.

The distribution of the internal force of the structure is shown in Fig. 9 and calculated by the structural analysis software. The results of the calculation show that most bamboo culms and joints reach design requirements in terms of the bearing

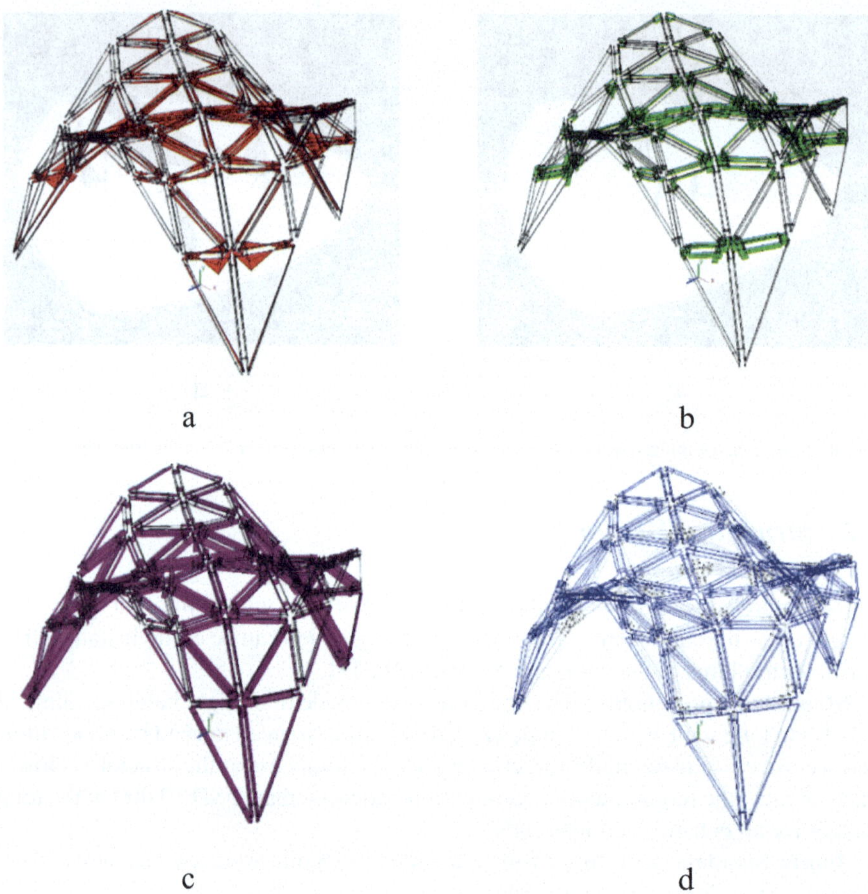

Fig. 9 Results of calculation of structural internal force and deformation: Bending moment (kN m) (**a**); shearing force (kN) (**b**); axial force (kN) (**c**); deformation (mm) (**d**)

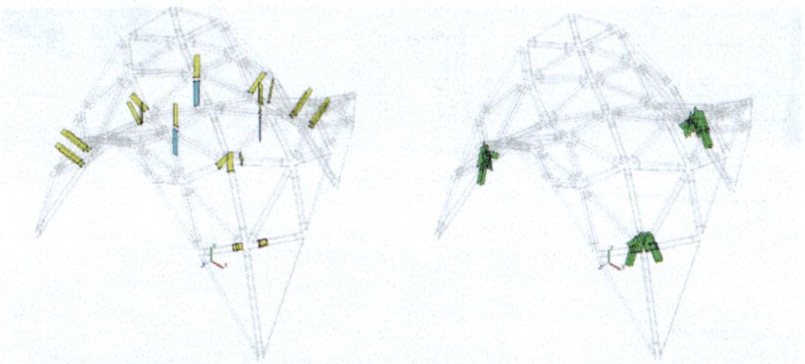

Fig. 10 Strengthened joints

capacity. For some connections with large stress, as shown in Fig. 10, the joints were strengthened by injecting cement mortar to improve the bearing capacity for the whole structure.

3 Construction

3.1 Triangle Unit Processing

Material Processing. Before the assembly, bamboo culms were treated by technologies of insect prevention, mold proof, and anti-corrosion to strengthen their durability.

Cut bamboo culms. According to the design, bamboo culms were cut to a certain length (Fig. 11) and two ends of each culm were cut at different angles.

Prefabricate triangle bamboo units. After bamboo culms were grouped, corners of each triangle unit were temporarily fixed by tape, and holes with 10 mm diameter for grouting were drilled in each corner of the triangle. Finally, according to the designed grouting length, cement mortar was injected into the corner node area with a pneumatic grouting equipment. As shown in Fig. 12, groups of prefabricated triangle bamboo units were prepared to be installed on-site in INBAR Pavilion.

3.2 Structural Assembly

In order to avoid the accumulated errors during the installation process and the negative impact of forced positioning on the structure, we adopted a more effective installation method to build the structure from the structural center to the edge, rather

Fig. 11 Cut bamboo culms

than traditional scaffolding or bottom-up installation. As shown in Fig. 13, the whole structure can be completed easily with the help of a tripod and a hand chain hoist.

Installation of the first block. There are 30 triangle units in the first block of the structures. First, assemble 6 central triangle units with temporary fasteners, and adjust their elevation, as shown in Fig. 14. Secondly, connect each two adjacent bamboo culms with bolts. Thirdly, loosen the fasteners and repeat the previous two steps to connect the remaining 24 triangle units until the first block was completed, as shown in Fig. 15.

Installation for the second block. There are 18 triangle units in the second block. After lift up the first block to the specified elevation with a tripod and a hand chain hoist, assemble the second block, as shown in Fig. 16a.

Installation for the third block. There are 6 triangle units in the third block. The installation method of the third block was the same as that of the second block (Fig. 16b).

Fix bamboo strips. Fix bamboo strips on bamboo culms in a criss-cross way, that could strengthen structural stiffness. Bamboo strips can become a substructure for the inner decorative layer (bamboo mats), as shown in Fig. 16c.

Fig. 12 Triangle units

Fig. 13 Installation method from the structural center to the edge

a b

Fig. 14 Central 6 units of the first block: Arrangement (**a**); elevation adjustment (**b**)

Fig. 15 Installation for the first block

3.3 Construction for Roof

The inner decorative layer of the bamboo mat, the waterproofing layer, the external decorative layer of thatch, and electric wires of light were installed, as shown in Fig. 16d.

Fig. 16 Second block installation (**a**); third block installation (**b**); fix bamboo strips (**c**); construction for roof (**d**)

4 Experience

We submitted our design to IBCC 2019 and built the first *Multi-culm Bamboo Spatial Lattice Structures* at INBAR Garden of the 2019 Beijing Expo. It took only two days to assemble the structure and complete roof decoration. This petal-shaped pavilion used the cheap raw bamboo culm as a structural member, while the triangle units were connected by ordinary bolts. Two expected goals of structural mechanical performance and installation efficiency have been achieved, and the feasibility and advancement of the structure have been proofed. Technologies of this new type of structural system can be spread to any country and region, because of the simple processing of structural components, low cost, easy purchase for joints, and good performance for the structure.

We summed up a few experiences and suggestions to guide further applications. Firstly, it is important to control the moisture content of bamboo culms, the mixing ratio of cement mortar, and the cutting accuracy of bamboo culm strictly. We should prevent bamboo culms from cracking to avoid shrinkage stress caused by the excessive shrinkage of bamboo culms in the later stage. Secondly, when using a tripod and a hand chain hoist to lift up the structure, we should pay attention to the number and the location of the binding points to prevent damage to the structure during the lifting process. If the structure becomes larger and higher, tripods and hoists cannot reach the lifting requirements and a crane needs to be used. Finally, before fixing the structure on the foundation, we have to pull ropes to connect the structure to the ground.

We hope this innovative system, *Multi-culm Bamboo Spatial Lattice Structures*, with the characteristics of factory manufacturing and assembly construction will have a better future. In order to achieve the goal of design and construct larger, stronger, and more beautiful bamboo structures, the Bamboo Spatial Structures R&D team of Zhejiang University will carry out more researches on morphology, mechanical analysis, and construction methods for bamboo structures in the future.

5 Renderings and Photos

Several renderings and photos of the pavilion as well as the photo of team members are provided below, shown in Figs. 17, 18, 19, 20 and 21.

Fig. 17 Renderings 1

Fig. 18 Photo 1

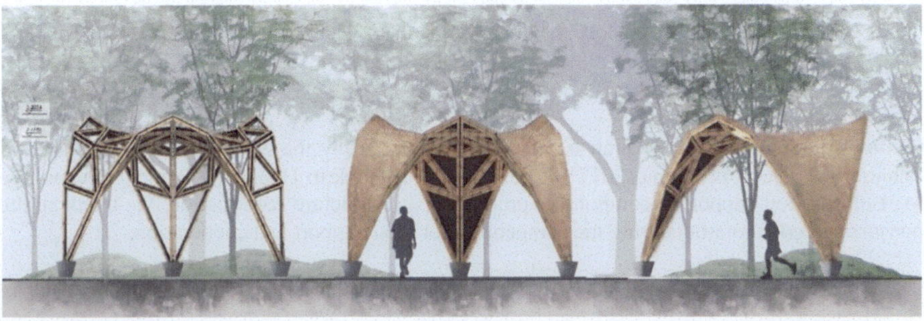

Fig. 19 Renderings 2

Fig. 20 Photo 2

Fig. 21 Team members from the left to right: Rui Guanbao, Wang Chidi, Zhuo Xin (advisor), Yang Chenchen, Huang Ting

Acknowledgements This research is financially supported by the National Major Science and Technology Projects of China (2017YFC0703500). We thank to Hangzhou Bangbo Technology Co., Ltd. for their support to components production and structure construction. We thank all the sponsors and collaborators during the competition for their support and cooperation.

References

1. Hong, C.K., H.T. Li, R. Lorenzo, G. Wu, H. Corbi, et al. 2019. Review on connections for original bamboo structures. *Journal of Renewable Materials* 7 (8): 713–730.
2. Zhuo, Xin, and Dong Shilin. 2019. Bamboo multi-culm spatial structure system and double loop fastener series, 201710025356. 4 [P]. Invention patent, March 8.
3. Albermani, F., G.Y. Goh, and S.L. Chan. 2007. Lightweight bamboo double layer grid system. *Engineering structures* 29 (7): 1499–1506.

Crouching· Hidden

Zhongzhong Zeng, Dingyi Liu, Yueqian Cai, and Xiaowen Qian

Abstract Crouching · Hidden is a building featured with traditional Chinese culture and is a tenon and mortise structure. In the famous movie "Crouching Tiger, Hidden Dragon", the toughness of bamboo is vividly demonstrated via the scene of two martial artists fighting on bamboo. The quality of bamboo represents the dialectical relationship between hardness and softness. Similarly, in terms of spatial organization, open space, and private space constitute two main parts of the structure. Following the scenes in the movie, we adopted the concept of letting people stand on pillars to realize the spatial function. Finally, a two-story building with an area of 9 m^2 was completed. Crouching · Hidden is such a fun space, where people can lie down and breathe fresh air, hide and seek, watch the scenery, do yoga, read books, chase, and play, take pictures, enjoy delicacies, etc.

Keyword Crouching · Hidden · Multifunctional space

1 Concept Design

Bamboo has been successfully applied to many designs as building material [1–3]. The work shows the dialectical relationship of space design on multiple levels while considering the structural design and path design. While reproducing movie scenes and showing Chinese culture, it also improves the functional attributes of bamboo buildings.

There is an old Chinese saying: "Stand high, see far". Raise the foothold to the height of the second floor, and you can see a good view. Therefore, we designed the building as of two floors (Fig. 1). Then we extended the building into a ring with an outer diameter of 9 m and an inner diameter of 6 m, because the circle is a gentle element that can express the flexibility of bamboo. Finally, to solve the traffic problem, we chose the ramp instead of vertical methods such as steps and ladders, because the ramp can ensure the continuity of the line of sight, allowing visitors to watch the surrounding scenery while walking, as shown in Fig. 2.

Z. Zeng · D. Liu (✉) · Y. Cai · X. Qian
School of Architecture and Design, Beijing Jiaotong University, Beijing, China

K. Liu et al. (eds.), *2019 International Bamboo Construction Competition*, Springer Tracts in Civil Engineering, https://doi.org/10.1007/978-3-030-91990-0_9

Fig. 1 Concept of design

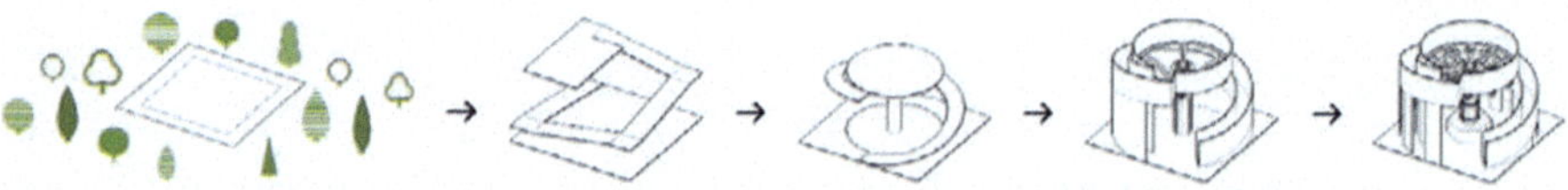

Fig. 2 Vertical transportation evolution

The title of the work is "Crouching · Hidden", which not only echoes the concept of the movie but also indicates two spaces with different attributes. The space of 9 m^2 is divided up and down: the public space on the upper floor faces nature without shelter, and all activities are carried out outdoors, which means "Crouching", indicating that there are no restrictions on the harmony between man and nature; the private space below, through the combination of the slab and the ramp structure, closes an inward space, which can be used for functional activities, which stands for "Hidden" and symbolizes the tranquility of the interior space.

Movement and stillness, Yin and Yang (in Chinese), opening and closing are integrated. As the time of the day goes by, and even the seasons of the year change, the two spaces with different attributes can meet the complex and diverse activity needs of different groups of people at different times (Fig. 3). With its virtual and actual features and opening form, this bamboo building can almost meet any needs

Fig. 3 Functional diversity in the time dimension

of tourists, such as yoga, reading, chasing and playing, resting, watching, taking pictures, dining, etc.

People of different personalities can also find a place that suits them here. Extroverts can stay on the open second floor, enclosed by railings, which has a transparent view. For outgoings, the second floor is a stage for display. For introverts, the first floor would be their preference. The first floor is surrounded by bamboo walls, leaving some curved windows. The natural and comfortable space is suitable for them to enjoy their time alone.

2 Design Development

This section introduces the process of generation and derivation of the design, and specific construction methods. This building is derived from the above-mentioned concept of "nature and blossom" and is entirely made of bamboo. The building covers an area of 9 m^2, with two inner and outer circles. The outer circle determines the area of the building, and the inner circle determines the area of the platform and the width of the ramp. The diameter of the outer ring is 9 ft, and the inner ring is obtained by advancing the outer ring 3 ft inward. The support of the entire building structure is calculated by mechanics to get the size of the central foundation, and the central column surrounded by 8 ft high bamboo poles is determined. The straight bamboo becomes the pillar, and the curved bamboo becomes the beam. The ideal bending and compressive resistance of bamboo, as a building material, is used to form a relatively stable structure. Its axial explosion diagram and structural details are shown in Fig. 4.

The section view of the axonometric diagram (Fig. 5) shows the relationship between bamboo and the earth clearly. The entire building is supported by a 2 ft concrete foundation from the ground. On the outer edge of the central column, bamboo is used to decorate the concrete foundation as circular seats in private spaces to meet people's needs for eating, reading, resting, or other behaviors that need to be completed in a sitting state. Since the entire second-story platform is supported by the central column, the single ramp will appear too thin. Thus, bamboo poles are used to enclose the inner ring and the outer ring at a certain interval while leaving an appropriate gap for people to walk through or play with children (Fig. 6). At the same time, the "wall" made of bamboo poles also provides a proper shielding of sight and a sense of enclosure for the first-floor space. The entire second-story platform has no other objects except the safety guardrail and the ramp entrance. The ground is paved with bamboo boards and waterproofed to prevent the building from being corroded by rain.

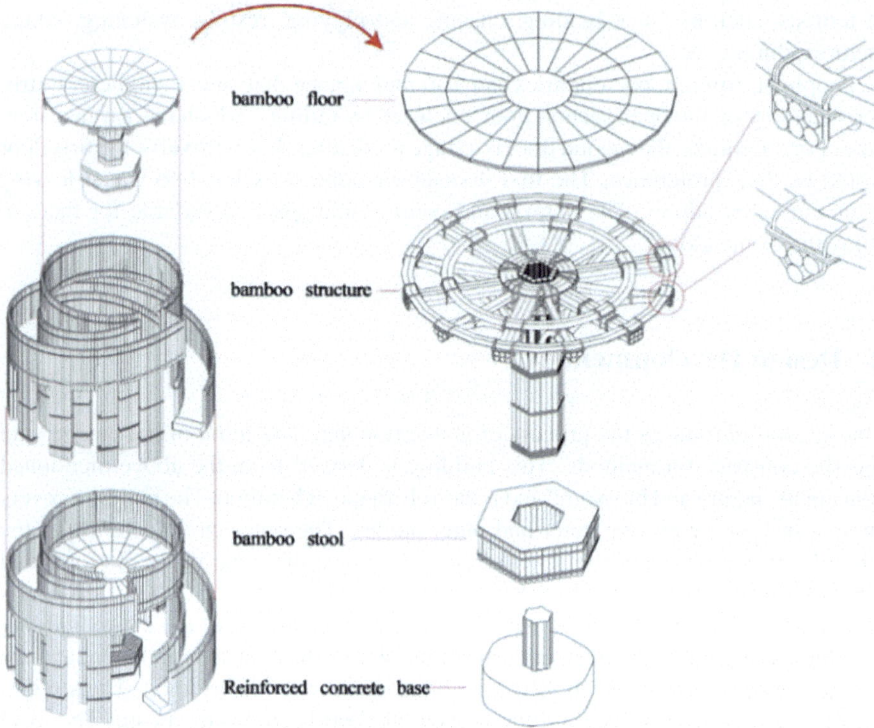

Fig. 4 Axial explosion diagram and structural details

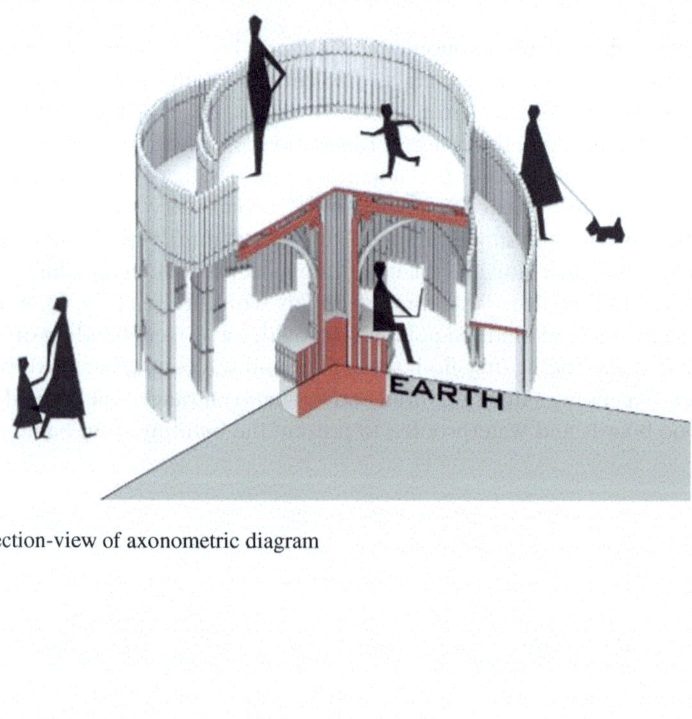

Fig. 5 Section-view of axonometric diagram

Fig. 6 Facade and usage of the gaps between structures

3 Experience

This section introduces the challenges and solutions faced by the design team in the design process, that is, bamboo as a building material is decorative and plastic, but it is difficult to meet the structural load-bearing requirements. To solve the problem, the design team took comprehensive consideration from three aspects, economy, beauty and structure. The final solution was to add components. In the competition, we learned that bamboo has infinite possibilities as a building material. Whether it is used as a structural material to support the entire building or as a decorative material to enrich and form a space, bamboo is a preferable choice. In the process of design and improvement, the vision was the focus of the project from beginning to end, and the two-tier structure had brought a lot of trouble to the team in subsequent design improvements. According to this concept, more people would stay on the platform of the second floor, which brought huge challenges in terms of structural strength. In the competition, the weight requirement "the structure can accommodate at least three adult men standing" could not be satisfied. According to our design, the entire structure should support about 10 people standing on the second floor. After mechanical calculations (Fig. 7), we had to add more bamboo as structural members in order to achieve this goal, which greatly increased the construction cost and made the project look bulkier.

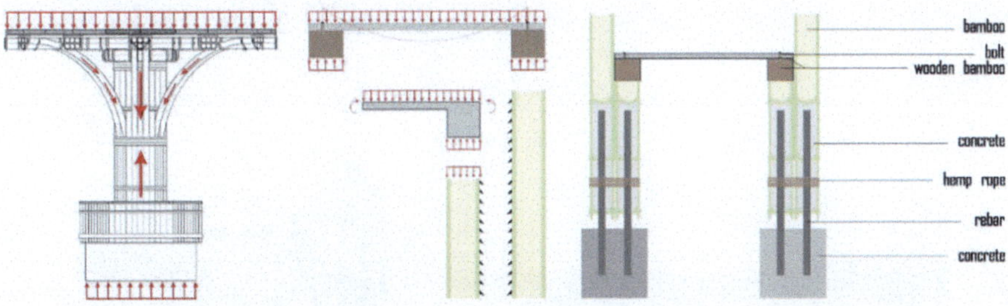

Fig. 7 Mechanical calculation and structure introduction

4 Renderings

This chapter mainly introduces the method of creating artistic conception during rendering. When rendering, as shown in Figs. 8, 9, 10 and 11, it is necessary to consider the lighting effects, the uniformity of colors, and the suitability of the

Fig. 8 "Crouching Tiger, Hidden Dragon" in urban space

Fig. 9 "Crouching Tiger, Hidden Dragon" in natural space

Fig. 10 "Crouching Tiger, Hidden Dragon" in different seasons

Fig. 11 "Crouching Tiger, Hidden Dragon" model photos

selected site to realize the natural integration of the building and the surrounding environment. A series of different scenes were produced, buildings were integrated into the urban space (Fig. 8), the natural environment (Fig. 9), different seasons (Fig. 10), and the night scene (Fig. 11) to create a feeling of artistic conception. In the process of creating and integrating into the city's artistic conception, the perspective shows the prosperity of the city, and the close-up shows the parks and green spaces in the city; in the process of creating and integrating into the natural artistic conception, we chose the lakeside scenery to show the harmonious artistic conception of architecture and natural ecology; in the performance of buildings in different seasons, representative scenes are also selected, with bamboo forests in summer and forests in winter as the background.

References

1. Zujian, Huang. Resource-driven sustainable bamboo construction in Asia-Pacific bamboo areas. 2021-07-15.
2. Xiang, Xingwei, Wu Qian, Zhang Ye, Zhu Bifeng, Wang Xiaoji, Wan Anping, Huang Tongle, and Hu Luoke. 2021. A pedagogical approach to incorporating the concept of sustainability into design-to-physical-construction teaching in introductory architectural design courses: A case study on a bamboo construction project. Sustainability 13 (14).
3. David, Witte. 2019. Structural bamboo building codes: catalysts for industry, research, and construction technology. *Technology|Architecture + Design* 3 (1).

Nomad Bamboo Pavilion

Liu Jin, Jindi Yang, Xin Yan, and Congcong Ren

Abstract Due to its environmental sustainability and high structural performances, bamboo is widely used in many engineering practices. This paper introduces a new bamboo-based design and construction system for the pavilion, which is named Nomad Bamboo Pavilion because of its portable and detachable features. The pavilion can be a moveable home, office, shop, etc. The main structure can be built by only one person with one screwdriver and the builder can use his/her hands to hang bamboo boards for walls, windows, and doors. The roof is constructed with the way of teeing round bamboo culms as tiles on the purlins. The pavilion can also be taken apart and packed up conveniently. Through modern architecture design, we bring the lost technique back and hope this construction system would give an example of the diversity of building technology to the industrialized world.

Keywords Nomad · "Lining and hanging" system · Interactive building · Bamboo pavilion

1 Concept Design

The Nomad Bamboo Pavilion (Fig. 1) is a multifunctional removable building, and it can be defined as a moveable home, office, shop, etc. The pavilion is composed

L. Jin · X. Yan
Center of Architecture Research and Design, University of Chinese Academy of Sciences, Beijing, China
e-mail: jinliu16@mails.ucas.ac.cn

X. Yan
e-mail: yanxin13@mails.ucas.ac.cn

J. Yang
Minmetals Land Limited, Beijing, China

C. Ren (✉)
School of Architecture and Urban Planning, Beijing University of Civil Engineering and Architecture, Beijing, China
e-mail: rencongcong@bucea.edu.cn

K. Liu et al. (eds.), *2019 International Bamboo Construction Competition*, Springer Tracts in Civil Engineering, https://doi.org/10.1007/978-3-030-91990-0_10

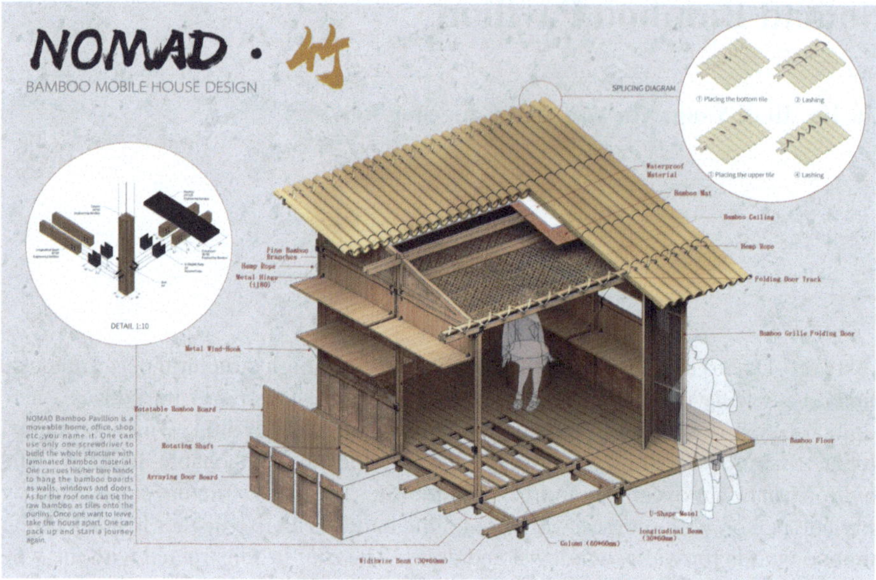

Fig. 1 Diagram for structural components of the Nomad Bamboo Pavilion

of three parts: a double-beam system, a "Lining and hanging" system and the roof. Bamboo is used to construct the main structure and metal for joints.

1.1 Materials and Construction

The double-beam system and engineered bamboo material products are used in the design (Fig. 2). Innovation has been a driving force behind expanding the uses of bamboo, particularly in China, where glued laminated bamboo products and production systems for specific bamboo species (usually *Phyllostachys Pubescens*) have been developed and contribute hugely to the nation's bamboo sector [1].

The industrialized production process can guarantee a precise position of the cavity for bolts in columns and beams, making them easy to be assembled. Using

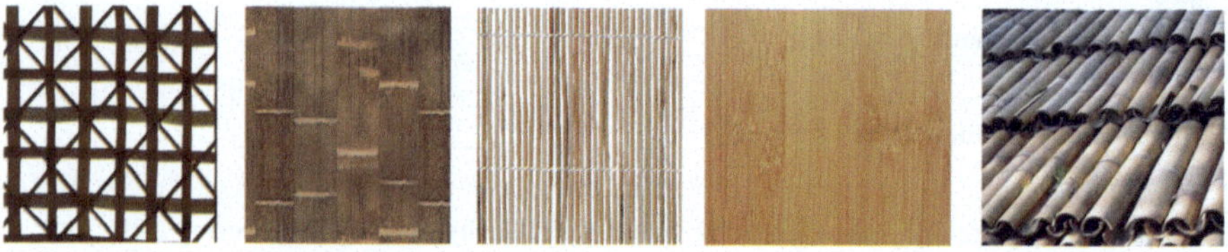

Fig. 2 Bamboo materials used in the design

Table 1 List of different components used in the pavilion

Bamboo products			Other products		
Name	Type	Size [mm]	Name	Type	Size [mm]
Bamboo tile	RB	150	Rope	Hemp rope	20/10
Bamboo mat	BP	/	Flexible waterproof	PE	0.1
Grille door	BP	300 × 2050 × 30	Hanging rail	Metal	33
Bamboo flooring	EBP	18	U-shaped parts	Metal	30 × 60 × 60
Bamboo column	EBP	60 × 60	Bolt	Metal	8
Bamboo beam	EBP	30 × 60	Rotating shaft	Metal	12
Lining door board	EBP	20	Wind-hook	Metal	6
Bamboo board	EBP	20	Hinge	Metal	44 × 33
Grille	RB	15	Hanging components	Metal	30 × 100
Bamboo weaving	BP	/			

Note RB—round bamboo; BP—bamboo product; EBP—engineered bamboo product

independent development of U-shaped metal connections, the double-beam system is "hanging" aside top columns by bolts. By adjusting the height of the foundation, the pavilion can adapt to any kind of topography.

Bamboo material is physically ductile and adaptable. Various bamboo materials produced by different processing can be used in different parts of the architecture. Bamboo is lightweight and it is difficult to fall because its flexibility makes it suitable for roofing skeletons in earthquake-prone areas [2]. Therefore, round bamboo tiles are adopted for roof of the pavilion. The structural components and walls of the pavilion adopt engineered bamboo as the main material, which provides high strength with a small cross-section. Bamboo boards, round bamboo with small diameter, bamboo grills are used for making walls and doors. Besides bamboo materials, prefabricated metal components are used to make connection nodes. The use of various bamboo materials creates a space hierarchy and makes the internal and external space of the house more exquisite and more refined. The list of different components use in the pavilion is provided in Table 1.

1.2 Lining and Hanging System

Lining doors, which use a mortise-tenon connection to hold each board horizontally, are widely used in Southern China (Fig. 3). When needed, boards can be removed to open the inner space, while it is fixed sequentially for solid facades. The beams for boards are very simple but effective. This type of buildings is popular for grocery

Fig. 3 Lining and hanging system inspired by architecture in southern China

stores in southern China. Users can either line or hang bamboo boards as walls or windows easily for enclosing or opening the house for daily uses.

The application of the "Lining and hanging system" (Fig. 4) is simple, convenient, and very suitable for small prefabricated buildings. Based on the concept, we made some adjustments and created a new lining and hanging system which can contain three changeable structure forms: hanging, supporting, and lining forms. The components used in the system are bamboo boards and simple prefabricated metal components, which are easy to install.

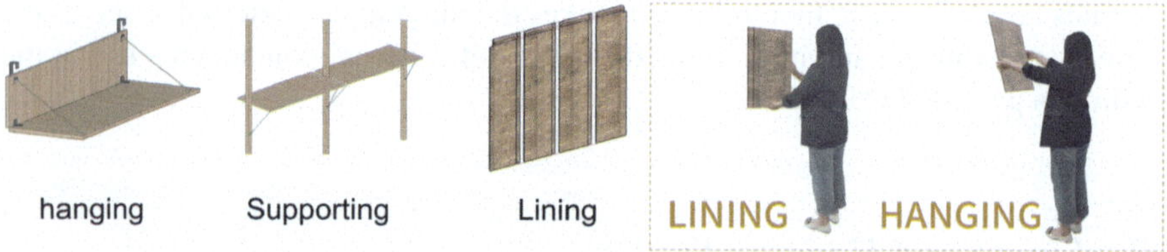

Fig. 4 Lining and hanging system

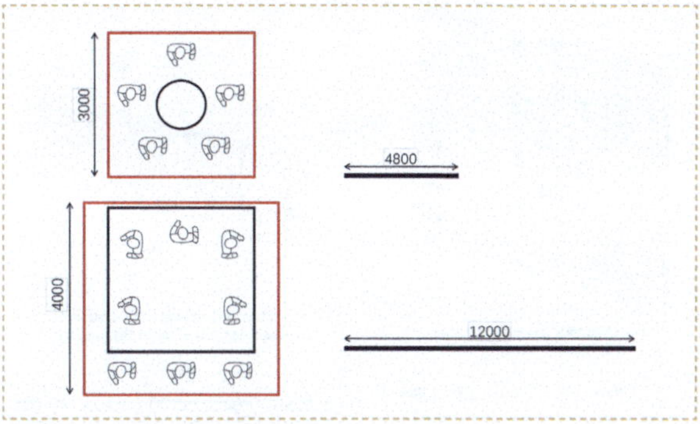

Fig. 5 Awning windows with different sizes

1.3 *Awning Windows*

The building area is very limited with only 9 m^2, so we adopt the strategy of turning the facade into furniture to increase usable area and improve its efficiency. In this case, we designed awning windows to save space (Fig. 5). Awning windows are parts of walls when not used. Once they are open, either towards outside or inside, they can be used as tables or awnings. During the hot summertime, the bamboo boards can be opened to provide shades and ventilation, while the bamboo boards can be closed under bad weather.

With all the ideas mentioned above, we encourage people to build their houses by themselves and adjust it towards changing environment and function requirements all by themselves. To become a nomad in modern society, anyone can get their own houses based on this strategy.

2 Scheme Design

2.1 *Lining System*

Based on the practice of "Lining door boards" of Jiangnan residential building components, the node optimization was carried out to consider physical characteristics of bamboo under the premise of retaining the basic principle. As shown in Fig. 6, each bamboo board forms a mortise-tenon connection through the left and right incisions. The door panel is applied to the back wall, and the left and right lower walls of the pavilion. In the whole wall, the gap between structural beams is used as the card groove to fix boards. Then the bamboo boards are connected from left and right in turn. The system is used for a whole wall and two lower walls of the pavilion. In this system, boards can be used as a table or sunshade by fixing with a wind-hook.

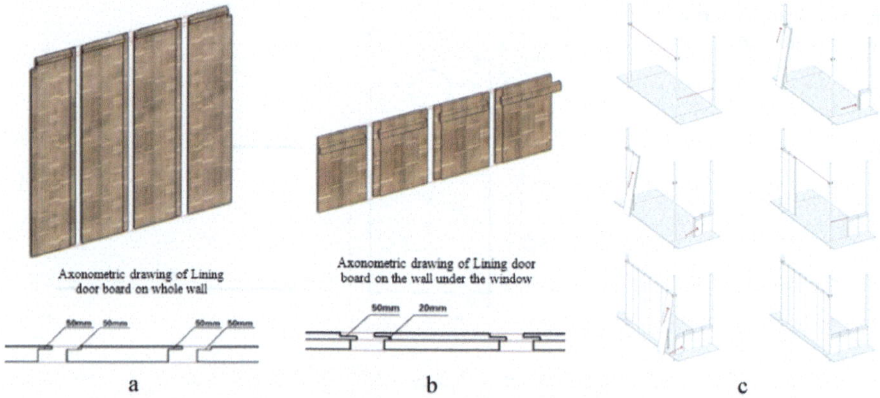

Fig. 6 Lining system. Plan of lining door boards on whole wall (**a**); plan of lining door boards on the wall under the window (**b**); schematic diagram for installation (**c**).

2.2 Hanging System

Considering its easy installation and detachable assembly, the upper bamboo board of facades uses a "hanging boards system" (Fig. 7). With the voids in the double

Fig. 7 Hanging system

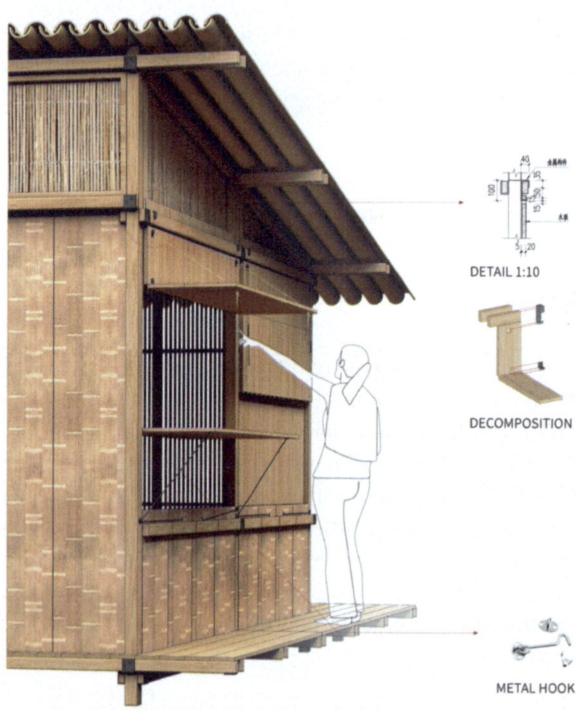

beams and prefabricated components, the bamboo board can be hung on the outer beam. The metal hinge is fixed in the middle of the bamboo board, which allows the bamboo board to be lifted. The lifted bamboo boards which can be fixed with a hemp rope play the effect of ventilation and shielding the sun.

The lower bamboo board is fixed with the column through the metal shaft, and metal hooks are placed on columns of bamboo boards on both sides. Through the shaft and the metal hook, the lower bamboo board can be set up as a desktop for exhibition, work, or other functions.

2.3 *Functions*

This design has strong extensibility because of the flexibility of structure, and it can be used in different ways, such as leisure, education, office, exhibition, and warehouse (Fig. 8). Depending on certain functions, the architecture form can be changed. When used as a warehouse, all facades are closed; When two sides are open, it can be used for exhibition; When it is an office, two sides of bamboo boards are open to provide more tables; When it is a leisure space, such as an area for drinking tea or coffee, the back wall can be opened to let people go through the space easily.

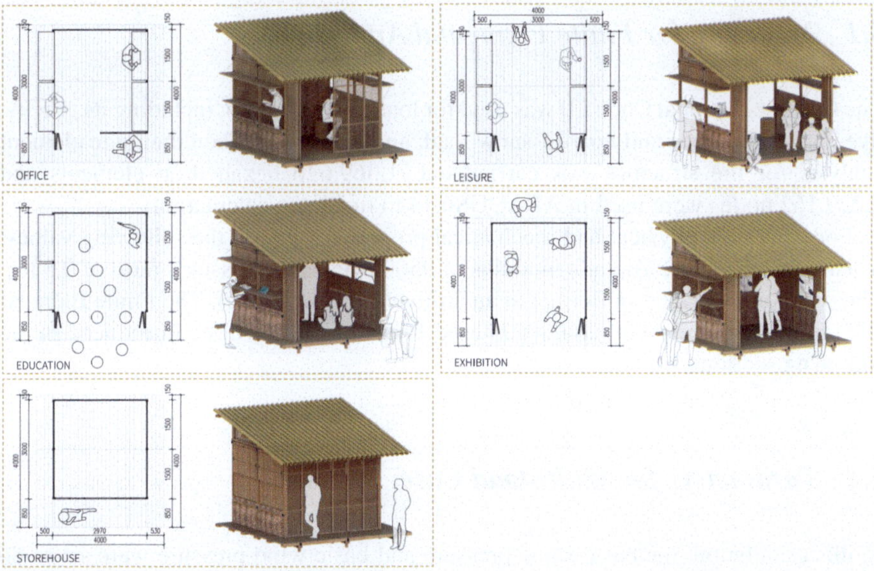

Fig. 8 With different functions

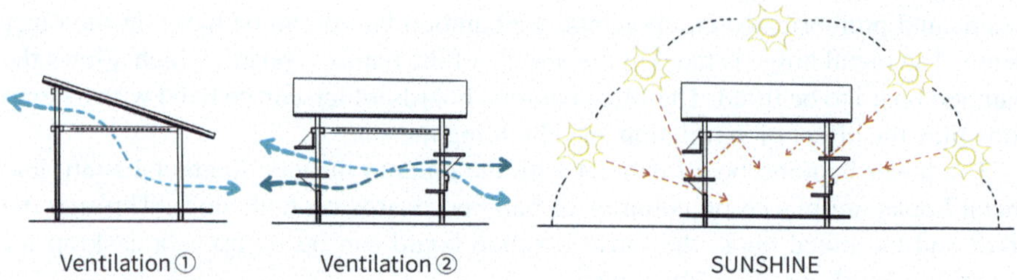

Fig. 9 Analysis for ventilation and sunlight

2.4 Analysis for Ventilation and Sunlight

As shown in Fig. 9, a bamboo grille ceiling is provided indoors to form a triangular cavity between the roof and ceiling, providing heat insulation and ventilation. Open bamboo boards installed on two sides can be used as a sunshade to avoid direct light and can facilitate natural ventilation during summer.

3 Structural Analysis

3.1 Modeling for Finite Elemental Analysis

The Chinese National Code [3] was used for load consideration, including its gravity, live load on the roof and boards, snow load, and wind load. Then the finite element analysis for the structure was carried out. 1,369,626 hexahedron elements and 2,229,162 nodes were used in ABAQUS6.14 to finish the calculation.

For bamboo's physical and mechanical properties, we use the following values: density of 1.5 g/cm^3, Young's modulus of 9500 MPa, and Poisson's ratio of 0.3 [4]. The pavilion is located in Beijing at an altitude of 54 m (MSL). The Finite Element (FE) model with relevant parameters used for calculation will be introduced in the following sections.

3.2 Parameters for Multi-load Case

In this calculation, the basic snow pressure and basic wind pressure were set to $S_0 = 0.45$ KN/m^2 and $W_0 = 0.50$ KN/m^2, which means once per one hundred years (Table. 2). Multi-load cases are implemented into the FE simulation to analyze the structural performance of the whole building.

The parameters of multi-load are set as below: the gravitational acceleration is set to 9.8 m/s^2; slab live load is set to 2.5KN/m^2; live load for the roof is set to 0.5KN/m^2.

Table 2 Parameters for loads in Beijing

Snow load (KN/m²)		Wind load (KN/m²)	
Level	Basic snow pressure	Level	Basic wind pressure
R = 10	0.25	R = 10	0.30
R = 50	0.40	R = 50	0.45
R = 100	0.45	R = 100	0.50

Snow load: the standard value for snow load was calculated as the following formula:

$$S_k = \mu_r \cdot S_0$$

in which S_k is the snow load standard value, and S_0 is the basic snow pressure. The roof snow distribution coefficient μ_r is set to 1.0 based on the code. So, the basic snow pressure for the structure is 0.45 KN/m^2.

Wind load: the wind load standard value was calculated as the following formula:

$$w_k = \beta_z \mu_s \mu_z w_0$$

in which β_z, μ_s, μ_z, w_0 are wind vibration coefficient, wind load shape coefficient, altitude variation coefficient of wind pressure and basic wind pressure, respectively. The wind vibration coefficient β_z can be calculated based on the following formula:

$$\beta_z = 1 + 2gI_{10}\left(\frac{z}{10}\right)^{-\alpha}$$

in which the peak factor g is set to 2.5, and the coefficient I_{10} is 0.39.

3.3 *Result Visualization and Analysis*

Gravity load case: The largest displacement is 6.145 mm. The largest first principal stress and smallest third principal stress are 4.632 MPa and −5.598 MPa, respectively, which meet the stiffness requirement of 8 MPa for tension and 17 MPa for compression (Fig. 10).

Slab live load case: The largest displacement is 1.616 mm. The largest first principal stress and smallest third principal stress are 2.868 MPa and −4.591 MPa respectively, which meet the stiffness requirement of 8 MPa for tension and 17 MPa for compression (Fig. 11).

a b c

Fig. 10 Gravity load case: Displacement (**a**); first principal stress (**b**); third principal stress (**c**)

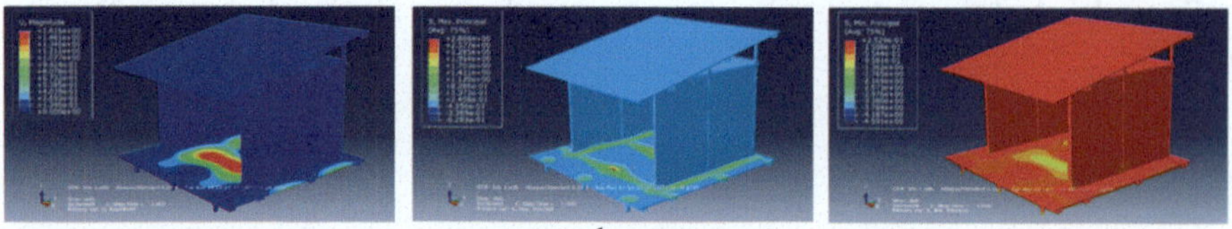
a b c

Fig. 11 Slab live load case: Displacement (**a**); first principal stress (**b**); third principal stress (**c**)

a b c

Fig. 12 Wind load case: Displacement (**a**); first principal stress (**b**); third principal stress (**c**)

Wind load case: The largest displacement is 13.35 mm. The largest first principal stress and smallest third principal stress are 7.185 MPa and −7.521 MPa respectively, which meet the stiffness requirement of 8 MPa for tension and 17 MPa for compression (Fig. 12).

4 Experience

We have been developing a space system for temporary use with the standard model of wood. In the course of this study, the material performance of bamboo was found to be better, and it happens to meet the bamboo architectural design competition, so the relevant design work is carried out. Due to the limited time of the competition, we didn't carry out experimental verification for nodes. We hope to continue to deepen and improve the design in the near future.

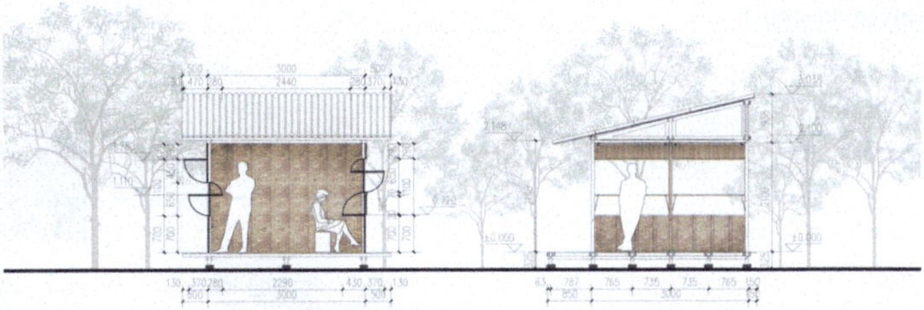

Fig. 13 Sections

Fig. 14 Elevations

Fig. 15 Interior space of the pavilion

5 Renderings and Photos

Several renderings (Figs. 13, 14 and 15) and photos of manual model (Figs. 16 and 17) for the Nomad Bamboo Pavilion are provided.

Fig. 16 Photo of Model 1

Fig. 17 Photo of Model 2

Acknowledgements This work is supported by National Natural Science Foundation of China (project number: 51908543). Authors gratefully acknowledge strong supports given by INBAR in the process of project design. We were greatly inspired by the research of Jiangnan residential houses, which made the design a success.

References

1. Benton, A. 2015. Priority species of bamboo. In *Bamboo: The plant and its uses. Tropical forestry*, vol. 10, ed. W. Liese, and M. Köhl, 32. Heidelberg: Springer.
2. Swathy Krishna, A.K. 2020. Review on bamboo as mainstream construction material. *International Journal for Research in Applied Science & Engineering Technology* 8 (XII): 794–795.
3. Chinese Load code for the design of building structures GB 50009-2012.
4. Janssen, Jules JA. 2012. *Mechanical properties of bamboo*, vol. 37. Springer Science & Business Media.

A Flower · A World

Sijia Zhao, Hongmei Zhang, Ziwei Zhang, and Xiangdong Zhao

Abstract The design of architecture begins at a teacup which can be opened and closed. Through a series of Chinese culture on "flowers", space narrative is carried out, and "bionic joints" are used for connections of the structure. By embracing the flower and the culture it represents, the design shows and metaphors the concepts of "Butterfly love flower", "Buddha sitting on the Lotus", and "Blooming flowers and full moon—Hua Hao Yue Yuan". In addition, the proposed project also combines the principles of ventilation (Joule–Thomson effect) and heat insulation (Bernoulli's principle). Finally, some consideration of ecological aspects is provided.

Keywords Bamboo · Architecture · Eco-ventilation · Air-conditioning · Butterfly love flower · Cultural symbol · Garden-style building

1 Concept Design

The building is surrounded by four bamboo walls with cave-shaped entrance and exit, and the internal stairway turns to the half-empty platform, like being placed in the mountains, as shown in Fig. 1. In the miniature world, changes of the visual plot and the light and shadow are rich, and the meaning of the building itself is the profound mountain forest. The building function is vague and polysemy, and the bottom layer can be provided with water points through the west opening.

Bamboo has been given a gentleman of high character since ancient times, such a gentleman's integrity has become the spiritual symbol of the Chinese nation and the precious wealth of Chinese history and culture [1]. Some elements of traditional Chinese culture are also used in the building. The architectural space absorbs the

S. Zhao (✉) · H. Zhang · X. Zhao
School of Architecture and Design, North China University of Technology, Beijing, China

X. Zhao
e-mail: 18610845192@163.com

Z. Zhang
School of Civil Engineering, North China University of Technology, Beijing, China

K. Liu et al. (eds.), *2019 International Bamboo Construction Competition*, Springer Tracts in Civil Engineering, https://doi.org/10.1007/978-3-030-91990-0_11

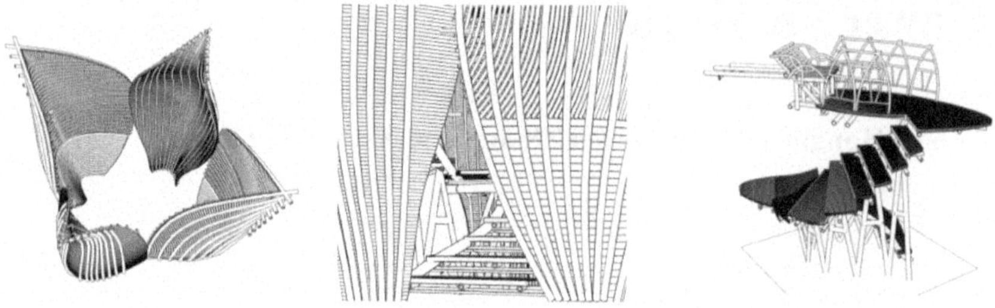

a b c

Fig. 1 Four bamboo walls (**a**); cave-shaped entrance and exit (**b**); internal stairway turns to the half-empty platform (**c**)

images of flowers and the cultural symbol they represent, which is wonderful between similarity and unlikeness.

The form of architecture sketches a teacup shown in Fig. 2 which can be opened and closed, that contains philosophies of heaven and earth, praise human education, and implies the metaphor of "Heaven and earth in pot" of Taoism.

The curved structure supports the butterfly-shaped abstract pattern, which is a wonderful display of the classic artistic symbol of "Butterfly love flower", and a reflection of the joyful heart in the world of mortals. The butterfly roof and the inner arch support are connected by bionic joints and clamped by exquisitely designed bamboo support, as shown in Fig. 3.

The top cover is arranged with three control chains divided evenly, which can be controlled in front and back according to the change of magnitude of sunshine. The cover could be rotated in a range of 10° from right to left, and of 30° from front to rear, as shown in Fig. 4.

The stairway turns into a two-story platform, and the lotus-shaped railings (Fig. 5a) surround a square inch of land. A seat to meditation for a moment is set up on the platform, which is a metaphor for "Buddha sitting on the lotus" (Fig. 5b). The slop

Fig. 2 Teacup

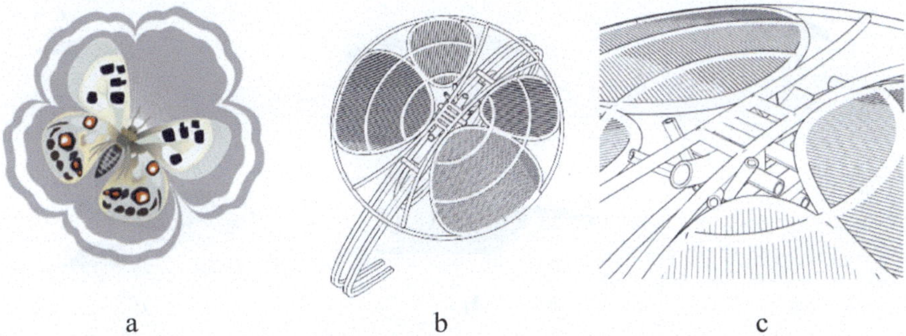

Fig. 3 "Butterfly love flower" (a); butterfly-shaped abstract pattern (b); bionic joints (c)

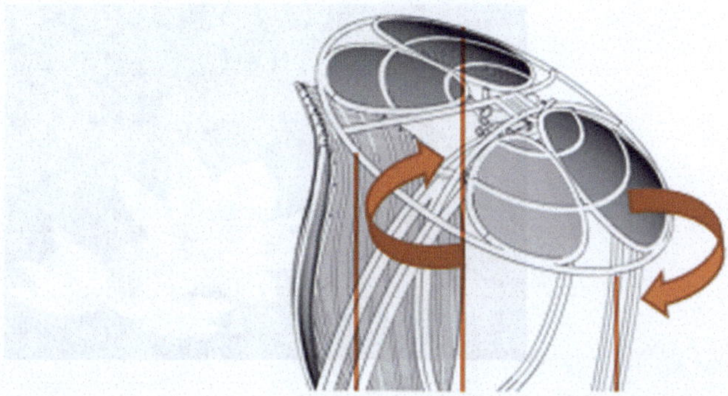

Fig. 4 Rotated cover

Fig. 5 Lotus-shaped railings (a); "Buddha sitting on the lotus" (b); slop of lotus platform (c)

of the lotus platform (Fig. 5c) is 1/9 to create a spatial illusion and to remind visitors not to stop for long.

The architectural structure reflects designers' thoughts. Successful works need to express an aesthetic sense based on physical structure [1]. The metaphor of the

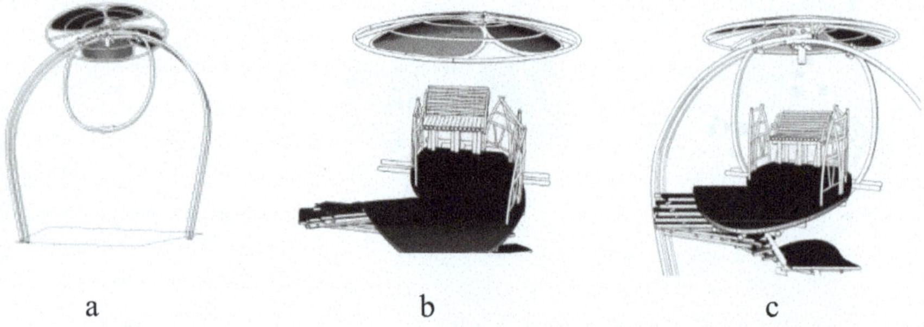

Fig. 6 Inner curved structure frame (a); seat (b); slop of the lotus (c)

Fig. 7 "Blooming flowers and full moon—Hua Hao Yue Yuan"

moon is formed by using the inner curved structure frame (Fig. 6). The combination of the design of the flower and the lotus table in the arc structure shows the traditional symbol image of "Blooming flowers and full moon—Hua Hao Yue Yuan" (Fig. 7), which means the beautiful and perfect life of the world.

2 Design Development

2.1 Building Placement

The building is placed in accordance with the dominant wind direction (see details in wind flow analysis chart) in summer and the roof is shaded, as shown in Fig. 8.

Fig. 8 Building placement

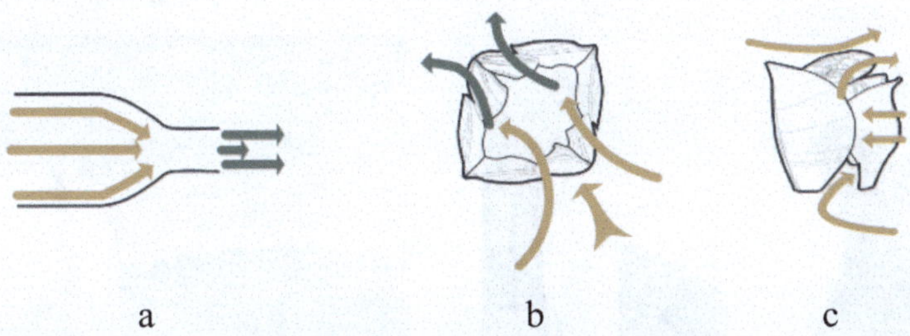

Fig. 9 Ventilation and cooling effect

2.2 Ventilation and Heat Insulation

The ventilation and cooling effects are achieved when a natural wind passes through the building. According to the Joule–Thomson effect, the velocity is accelerated and the pressure is decreased by the narrow space between the bamboo walls (Fig. 9a, b).

The bamboo cavity is naturally insulated and has a few overheads at the bottom of the building, forming a typical chimney-shaped shaft inside. If the wind blows overhead, according to Bernoulli's principle, the patio further accelerates the internal airflow, taking away heat, to realize ventilation and cooling. At the same time, due to the vertical temperature difference caused by sunshine, the hot air rises and drives the air below to replenish upward, which is conducive to ventilation (Fig. 9c).

2.3 Functional Scenario

Several functional scenarios are provided for the design, as shown in Fig. 10.

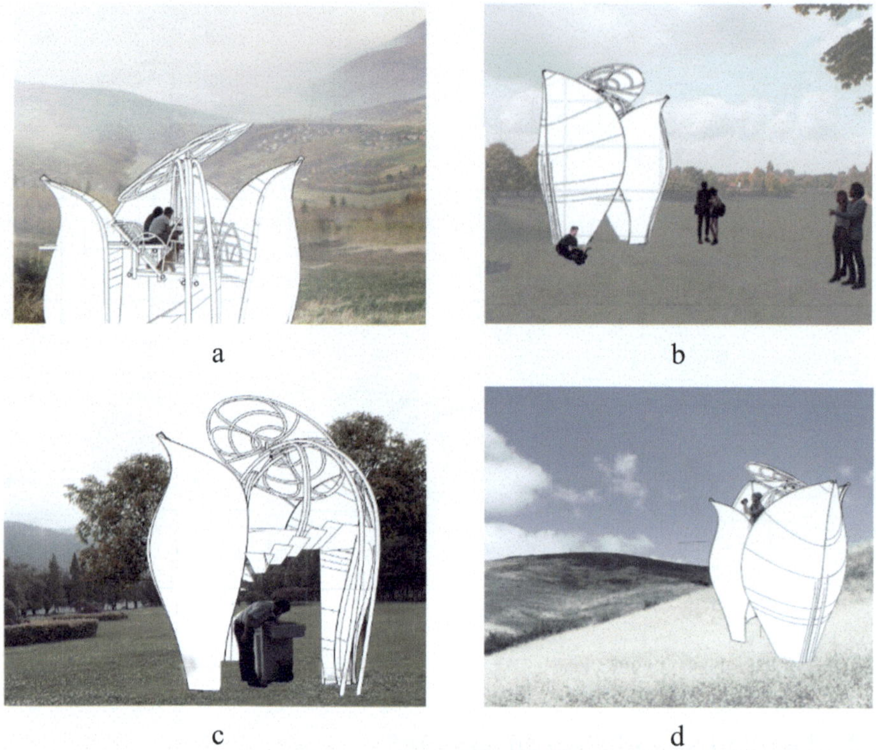

Fig. 10 Rest station (**a**); scenery in the park (**b**); Fountain (**c**); platform of overlook (**d**)

Fig. 11 Land of Peach Blossoms

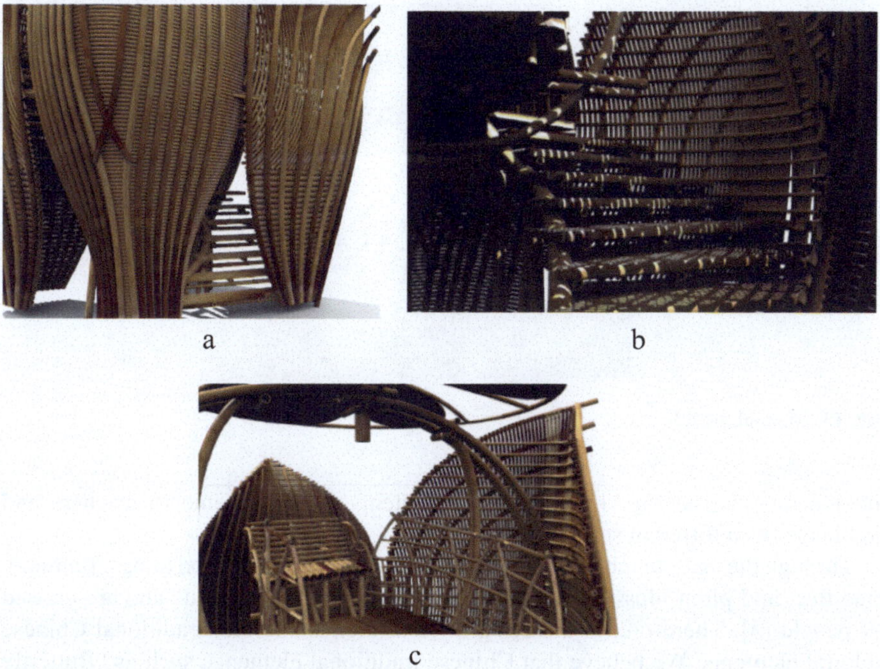

Fig. 12 Entrance (**a**); Stairs (**b**); the top (**c**)

2.4 *Space Narrative*

Our design is based on the narrative of the story of Land of Peach Blossom (Fig. 11), which is a classic, telling the story of a man who strayed into a wonderland. He went further to explore, and when he came to the end of the grove, he saw a spring which came from a cave in the hill. Having noticed that there seemed to be a weak light in the cave, he tied up his boat and decided to go in and explore. At first, the opening was very narrow, barely wide enough for one person to go in. After a dozen steps, it opened into a flood of light. He saw a wide, level valley, with houses, fields, and farms [2].

As shown in Fig. 12, visitors to our design could also have the same feeling: they see narrow steps, then go upstairs, and finally find an open platform for overlooking beautiful scenery.

3 Discussion

Three members of our team went to the 2019 Beijing International Horticultural Exposition to participate in the competition. We introduced the design concept,

Fig. 13 Manual model

manual model (see Fig. 13), and other contents of the scheme to teachers and architects from different schools.

Through the ages, bamboo has been regarded as a symbol of good things. Bamboo, pine tree, and plum blossom are called the "Three friends in cold" and are praised by people [3]. Therefore, in the design, we referenced several traditional Chinese cultural elements. We believe that Chinese traditional elements, such as "Butterfly love flower", could make our design outstanding. However, there is a problem of unstable support force on the structure which we could not solve and we still need further consideration to improve the whole supporting structure. In a word, we saw the excellence of many other teams, and we also had some communication with many teachers and architects, which improved our architectural quality a lot.

4 Renderings

Two renderings of the whole structure are provided, as shown in Fig. 14. Some lights are added inside the rendered model, where the structure supported by bamboo is clearly visible (Fig. 14a), while the exterior is surrounded by bamboo walls that separate the interior from the exterior (Fig. 14b). People can feel the artistic conception of their own little world of heaven and earth in nature.

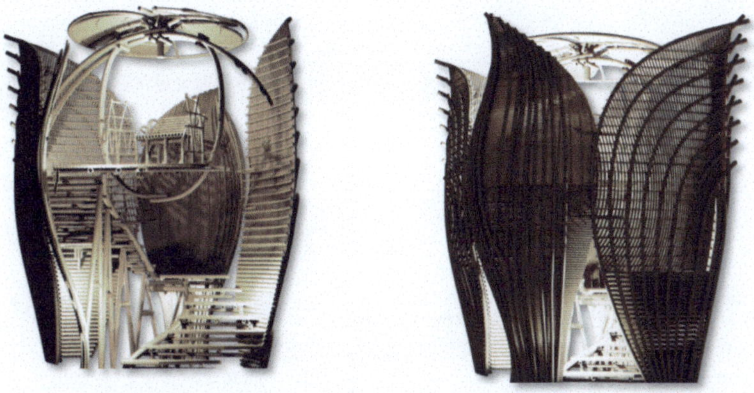

a b

Fig. 14 Renderings of the whole structure

References

1. Xuan, Y.K., and S.S. Li, 2017. *Research on the application of bamboo materials in gardens and architecture*. Chengdu University of Science, School of Tourism and Urban-Rural Planning (in Chinese).
2. Tao, Y.M. Origin of Peach Blossom. Webpage: https://hanyu.baidu.com/shici/dtail?pid=f3f889357df-84af681cfc55098ced685&from=kg0.
3. Wang, X.W. 2017. *Bamboo and architecture*. China Aerospace Science and Technology Group 7 (in Chinese).

Bamboo Field

Qing Gong, Yin Zhang, Ruonan Qin, and Kewei Liu

Abstract The design is inspired by the ancient rattan ball for sports. Through researches on its material and structure, the beautiful shape and solid characteristics of the rattan ball help us deepen the design. Combined with traditional Chinese bamboo weaving technologies, a simple and stable structure is designed. The design uses the good flexibility of bamboo to form a spherical grid structure. Woven bamboo strips restrict each other to ensure the firmness of the structure. The spherical grid structure allows the air in and out, with changing light and shadow. A membrane is used to shield sun and rain. Structural analysis was also carried out to check the deformation of the structure and the stress strength of components.

Keywords Spherical grid structure · Rattan ball for sports · Bamboo weaving technologies

1 Concept Design

Bamboo, as a fast-growing renewable material, is expected to be a sustainable alternative for some traditional structural materials, such as concrete, steel, and timber [1]. Meanwhile, bamboo is the main structural material required by the competition. Combined with the theme for which we want to express, "build a house in a bustling settlement but I can still not be bothered by the clatter of the traffic, and keep a heart away from the worldly world is the secret of my peace", we carried out the design. The concept design of our pavilion is to be introduced from the following three aspects: spherical grid structure, rattan ball for sports and bamboo weaving technologies.

Q. Gong · Y. Zhang · R. Qin
Department of Architecture, Xiamen University, Fujian, China

K. Liu (✉)
International Bamboo and Rattan Organisation, Beijing, China
e-mail: kwliu@inbar.int

School of Civil Engineering, Tsinghua University, Beijing, China

K. Liu et al. (eds.), *2019 International Bamboo Construction Competition*, Springer Tracts in Civil Engineering, https://doi.org/10.1007/978-3-030-91990-0_12

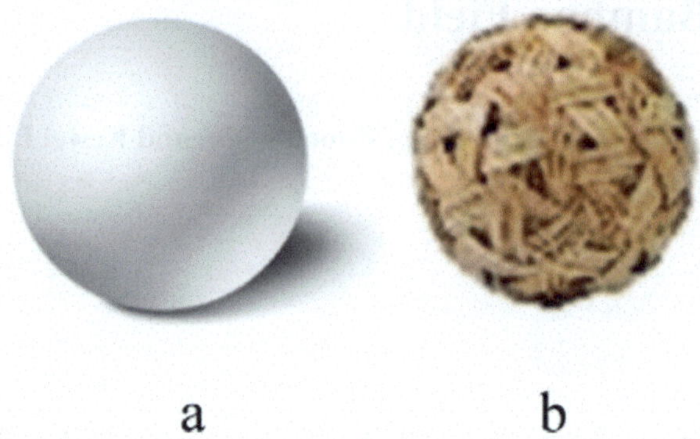

Fig. 1 Sphere (**a**); Rattan ball for sports (**b**)

1.1 Spherical Grid Structure

The sphere (Fig. 1a) is a very solid structure that gives people a feeling of comfortable and pure that is consistent with one of the targeted functions of our pavilion. The purpose of our design is to explore how to construct a spherical grid structure by using bamboo. The characteristics of bamboo could be also fully applied to the structure at the same time.

1.2 Rattan Ball for Sports

Based on the idea of constructing a spherical structure, an ancient Chinese rattan ball for sports came to our eyes after studying a series of different materials. To imitate the structure of the rattan ball became the object of our design and research. The rattan ball is a Chinese ancient handicraft, which is easy to make and its structure is very stable. It was introduced to Southeast Asia and became a popular sport in the 1940s, known as "sepak takraw" (in Malaysian) [2]. The ball is woven by rattan strips, with uniform stress and gaps between different strips, as shown in Fig. 1b. We can imagine that the rattan ball is a building with a good structural performance, and it also meets the ventilation and light requirements of the building.

1.3 Bamboo Weaving Technologies

Traditional bamboo weaving technology has a long history and bamboo woven products are closely related to people's daily life. Weaving is also one of the basic forms of construction in ancient times [3]. Therefore, on the one hand, we aim to use traditional weaving skills to reflect the good characteristics of bamboo in a small-scale structure. On the other hand, we also hope to spread this great Chinese craft to the

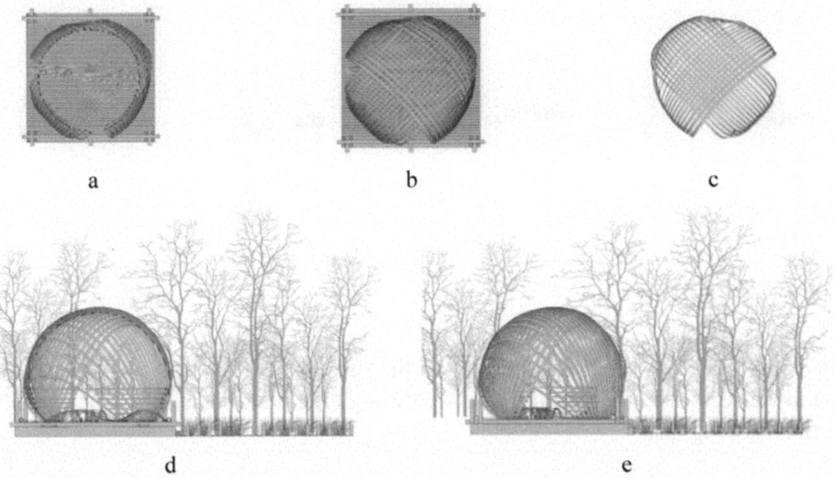

Fig. 2 Plan (**a**); Top view (**b**); Upward view (**c**); East elevation (**d**); West elevation (**e**)

world. The biggest challenge for us is how to scale up a small woven object into a real functional structure.

2 Design Development

2.1 *Profile of the Pavilion*

The planned building area is about 7.06 m^2 and the whole site area is 9 m^2. The plan of the structure, its top and upward views, as well as its elevations, are shown in Fig. 2. The structure is assembled by prefabricated bamboo strips through metal connections. We consider to use a membrane to form an external façade and use round bamboo culms to pave the ground.

2.2 *Function Analysis*

This structure can achieve two functions: if we do not cover any material outside of the bamboo structure, then the pavilion can become a completely open leisure space; if we tie the membrane to the bamboo strips, the pavilion can be rain-proof, sun shading and with other functions.

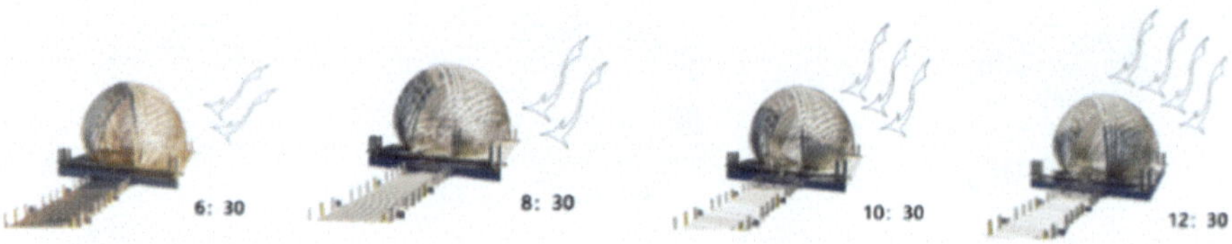

Fig. 3 Sunlight analysis chart

2.3 Light Analysis

By using bamboo weaving technologies, the structure can present different light and shadow effects when the sunlight passes through the building at different time points of the day (Fig. 3).

2.4 Processing of Construction

After studying a series of references, we plan to construct it in several steps as follows: Firstly, choose bamboo species with good toughness, such as Moso bamboo, then split the whole bamboo culm into strips with certain width; Secondly, weaving bamboo strips to form a spherical grid structure; Thirdly, insert ends of bamboo strips into four foundations and pour concrete to fix them.

The structure is smooth and continuous with coherent external lines to form a space with shelter function, presenting bamboo's natural tension and beauty. The space can subtly reflect the color of the sky as time goes on. The sunshine in the day and the light at night create delicate light and shadow. The external membrane makes the edge of the bamboo structure blurred, presenting the flexibility and permeability of the structure, as well as the beauty of bamboo's soft and rigid.

2.5 Nodes

In the construction of modern bamboo structures, metal bolts are usually used for connections. However, to match bamboo's natural simplicity, we choose the way of lashing rather than bolts, while maintaining the organic overall characteristics of the structure. Lashing by hemp rope (Fig. 4) not only highlights the natural characteristics of bamboo architecture but also inherits the aesthetics of traditional bamboo construction technologies.

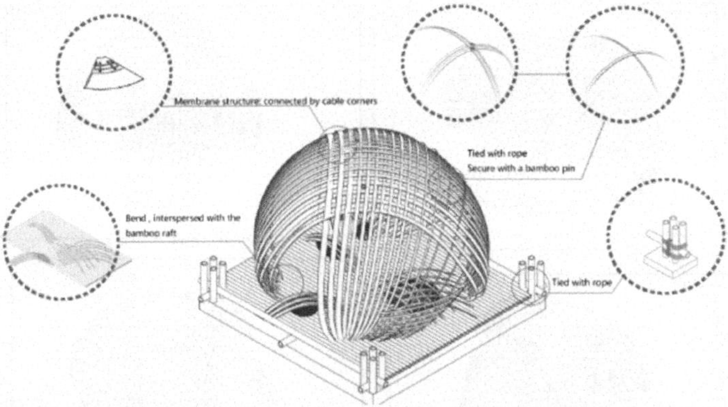

Fig. 4 Nodes and other details

2.6 *Modeling and Analysis for Structure*

Bamboo has high tensile strength along the fibre, which is about 2 times of wood. Its compressive strength is about 1.5–2 times of wood, and its tensile strength per unit weight is about 3–4 times of steel. Therefore, bamboo is a very good structural material. Through weaving technologies, bamboo can create curved structures with large spans.

The parameters of bamboo used in the calculation are shown as follows: the density of bamboo is 800 kg/m^3; the modulus of elasticity is 9500 MPa and the Poisson's ratio is 0.3. The cross-sectional area of bamboo strips was 50 mm $\times$ 9 mm. We used Ansys to analyze the structure and the results are shown in the following paragraphs. ANSYS software is a large general finite element analysis software. Can carry out structural analysis, static analysis, load analysis and so on.

Deformation: According to Fig. 5, the deformation of the structure in the X, Y, Z directions are -0.018 m~0.019 m, -0.021 m~0.01 m, and -0.025 m~0.002 m respectively. Deformations in three directions are all within the controllable range, which satisfy the deformation control conditions.

Strength: the maximum compressive stress, the maximum tensile stress, and the maximum shear stress of the structure are 7.79 MPa, 10.1 MPa, and 1.7 MPa respectively, which are lower than the allowable stress of selected bamboo materials.

3 Experience

IBCC 2019 aims to invite global students who are learning architecture, civil engineering, landscape, and other related majors to show the potentials of bamboo as a construction material. Thus, we had such a wonderful opportunity to compete with many other students from around the world. During the competition, we learned a

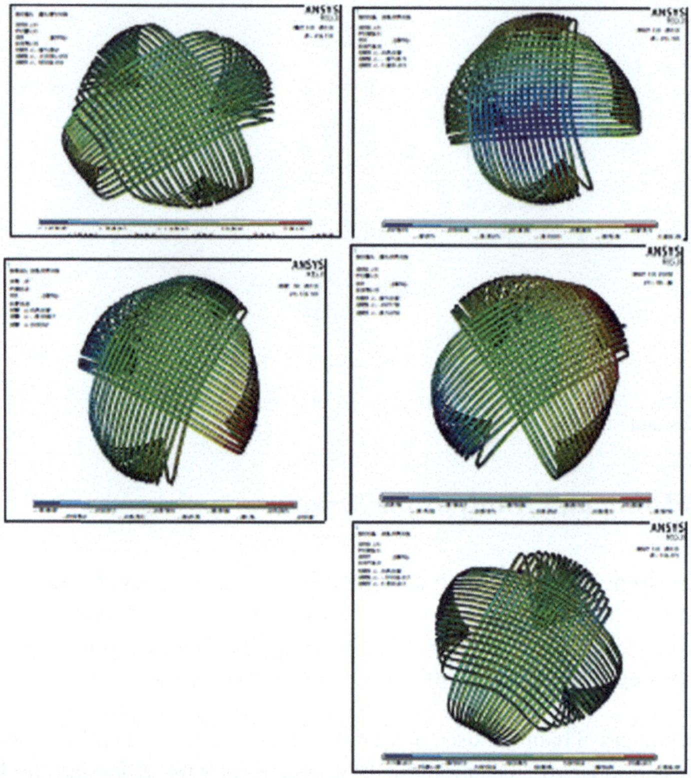

Fig. 5 Structural analysis diagram

lot which we wouldn't normally touch with. Our team, consisting of three students, Yin Zhang, Qing Gong, and Runan Qin, from the architecture department, was led by Mr. Lin Yuxin. From the initial design to the final submission, all the members cooperated very well. We learned how to help each other and how to overcome the difficulties, which were extremely valuable experiences for us.

3.1 Preparation for Presentation in Beijing

One month after submitting the design, we received the good news that our project was shortlisted as one of the top 15 finalists. This result was the best reward for our efforts in the past several months. However, we cannot relax, because the final result was to be decided during the Jury Review Meeting in the coming month in Beijing. Therefore, do our best to prepare the presentation had become a top priority for us. As time was so limited, we almost spent all of our spare time on the discussion of how to show our design concept as well as how to prepare a small model required by the Organizing Committee. As shown in Fig. 6 of the construction flow chart, we made a 1:50 scale model, for which we used hemp rope to strengthen the stability of the structure. Meanwhile, we purchased some little lamps with knitting elements to decorate it. After several days and nights, a realistic model (Fig. 7) was created.

Finally, Zhang Yin and Qin Ruonan represented the team to go to Beijing, and Gong Qing gave supports from Xiamen. When we spoke English fluently to introduce our design to the jury, we were very excited and felt so proud of ourselves.

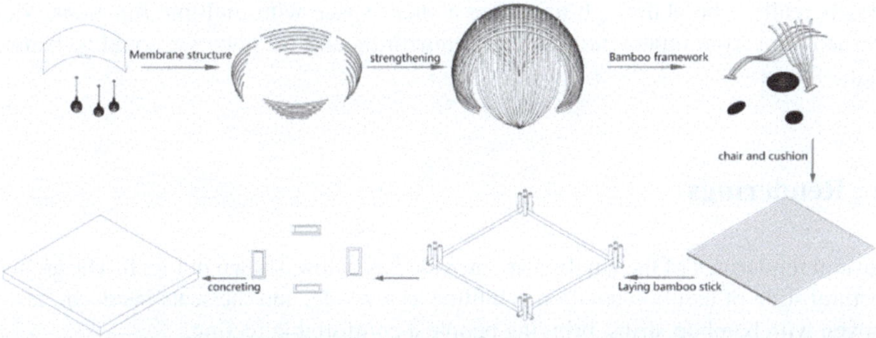

Fig. 6 Construction flow chart for model

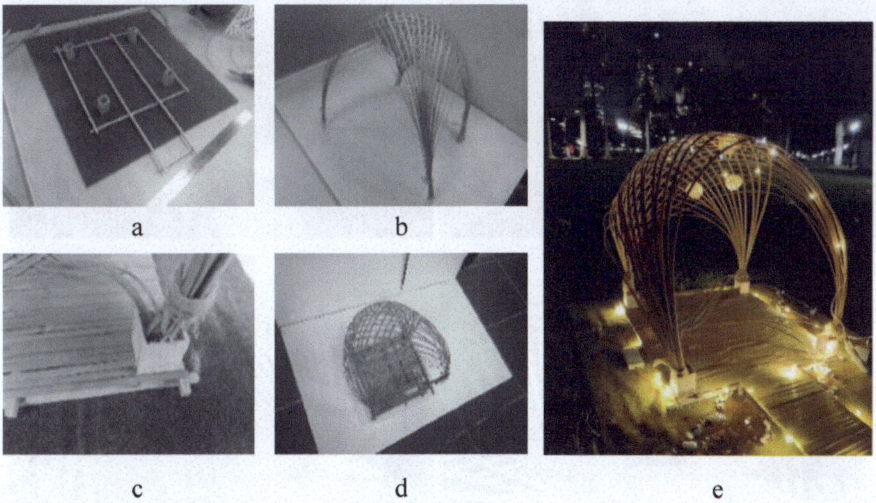

Fig. 7 Photo of the model

3.2 *Ideas for the Future*

For this design, we paid too much attention to the concept and pavilion's shape, but we didn't give enough consideration for bamboo's performances, such as whether bamboo could really help us to complete the complex weaving to form a building, or whether the space formed by bamboo can accommodate more activities. In Beijing, we saw four structures in 1:1 scale at INBAR garden which were completed by other teams. One of them even accomplished a two-floor structure, for which we think their idea is really a breakthrough to realize a small space with multiple functions. We learned a lot from others through this competition and we hope we could be better in the future.

4 Renderings

Several renderings of the pavilion are provided as below, shown in Fig. 8. The architectural style of bamboo pavilion is unified as a whole, and the seats inside are also woven with bamboo strips, bringing people a comfortable feeling.

Fig. 8 Renderings: Side perspective (**a**, **b**); Interior space (**c**, **d**)

Acknowledgements Our participation in IBCC 2019 was well guided by our instructor, Mr. Lin Yuxin. His patient guidance with a rigorous and scientific attitude made our design moving forward on the right path and got a final success. Thanks to Xiamen University providing the budget to support us. Finally, we would like to deliver our appreciation to the organizers of IBCC 2019 who gave global students a chance to brainstorm together and provide us a big stage to show ourselves and compete with others.

References

1. van der Lugt, P., A.A.J.F. van den Dobbelsteen, and J.J.A. Janssen. 2006. An environmental, economic and practical assessment of bamboo as a building material for supporting structures. *Construction and Building Materials* 20 (9): 648–656.
2. https://en.wikipedia.org/wiki/Sepak_takraw. Last accessed on Aug 2021.
3. Gao, J. 2011. Ordinary bamboo works. China Academy of Art.

Push and Pull, Varying Tranquility

Sihan Wang, Yu Wang, Ruojun Ren, and Fei Wang

Abstract "Push and Pull, Varying Tranquility" is a tea house designed for the 2019 International Bamboo Construction Competition (IBCC 2019). Its form and inner space can be changed by users. We designed the appearance and decoration in light of the characteristics of bamboo. We also want to enhance the relationship between humans and nature by stimulating interactions. To achieve that, we designed a special component and made it to become a part of the building. The space has multiple definitions by using its transparency characteristic to create different experiences and expand the overall space of building in human sensation.

Keywords Bamboo · Variable architecture · Green building

1 Concept Design

Being in harmony has always been an important topic in the field of architecture. The use of natural materials is environmentally friendly which has important social and environmental value. What role natural material plays and how it can inspire people, that are the questions we concern about. Nowadays, participation as an activity has drawn more and more attention from designers and artists [1–4]. People are no longer just viewers in an exhibition, but a part of their artwork. By getting people involved in, artworks and architecture can change their form and appearance, and gain new meanings as well. We want to create opportunities for this kind of involvement and stimulate interactions between people and the building, not only to make people talking to the material which we used but bring them to a new and incredible experience.

S. Wang (✉) · Y. Wang · R. Ren · F. Wang
School of Architecture and Civil Engineering, Xiamen University, Xiamen, China

K. Liu et al. (eds.), *2019 International Bamboo Construction Competition*, Springer Tracts in Civil Engineering, https://doi.org/10.1007/978-3-030-91990-0_13

Under the thinking of these ideas, space and materials become the main focuses in this design.

In the aspect of space, we try to create multiple spaces by transparency characteristics, therefore people can experience more than one space in one place.

In the aspect of material, we aim to guide people to experience, understand and appreciate the material by creating a tranquil atmosphere and stimulating interactions. The material we use is bamboo, which has a cylinder shape and slender form as its natural features, and has connotations as one of the "three cold weather friends",[1] and is often praised by Chinese people. Bamboo's natural features and connotations give us the inspiration and ideas of the whole design. We made push–pull sliding doors, the open and close of push–pull sliding parts changes the spatial relationship of the building. The appearance is designed as pure and neat as possible, the distinctive uprightness and hollow-hearted characteristics of bamboo, and vertical arrays of bamboo poles and sheets are used as skin to show the form of bamboo.

We aim to create an interesting connection between architecture and its users, so we designed different spaces with multiple definitions and made them all infiltrated into each other. The architecture has a changeable appearance and special push–pull sliding components. Light and shadow create an incredible experience which also gives users a chance to trigger the change of space.

2 Design Development

2.1 Form Generation

The inspiration of the architectural form is from bamboo which grows in nature. We abstracted the sharp corners from the shape of the bamboo shoot as which of building's roof, and designed two roofs that gradually rise up, symbolizing the blooming and vigorous growth of life. In the design of the building facade and structural form, we adopted the image of bamboo growing vertically in the forest. Part of the wall was designed as a vertical arrangement of bamboo materials to show the vitality of the building. For the design of internal space, we divided the mass into blocks based on its height and created a comparison between empty space and solid masses. Then we added a horizontal structure to form a cubic space, which is nested with other space. Finally, we divided the interior space of the building further by bamboo grill, enriching space layers, and positioning sliding doors (Fig. 1).

[1] "Three cold weather friends": pine, bamboo, and plum blossom. Because these three plants can still maintain tenacious vitality in cold winter, they are the symbol of noble personality in Chinese traditional culture.

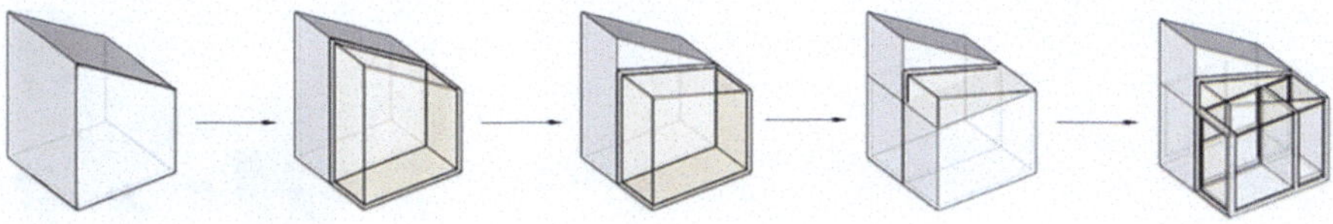

Fig. 1 Form generation

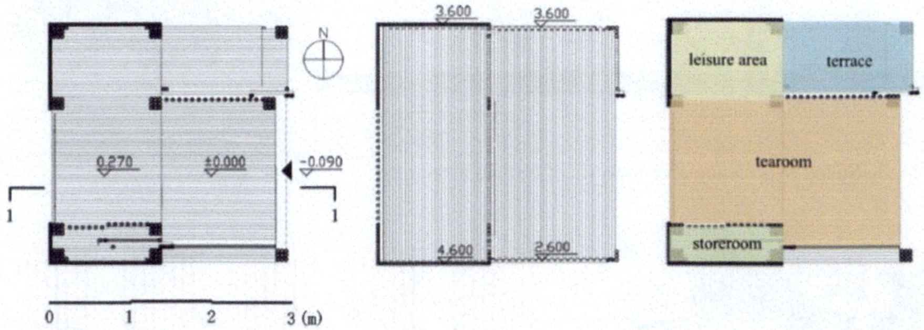

Fig. 2 Floor plan and functional partition

2.2 *Floor Plan and Functional Partition*

The plan of the building is a regular square. The floor is raised to form a platform where people can stay, sit and enjoy the scenery (Fig. 2). Due to the interactivity and spatial variability of the building, the activities in the space become more varied and enriching. When the sliding doors are not pulled out, the tea house is open to the landscape completely, which could show a clear view. When the doors are pulled out, the tea house is bathed in the light and shadow of the bamboo grill, and the space also becomes more private and quieter than before. In addition, the north and south sides of the building are divided by sliding doors, creating some small space for storage, viewing, and any other purposes (Fig. 3).

2.3 *Development of Push–Pull Sliding Component*

The change of the spatial interface is achieved through the sliding doors. Considering the human scale and building scale, we finally chose the metal framework of the foldable sofa bed as the inspiration for the design (Fig. 4). Besides, the framework of the door is designed with a uniform spacing and extra space, in case the undulating surface of bamboo ensures smooth sliding and safety [5].

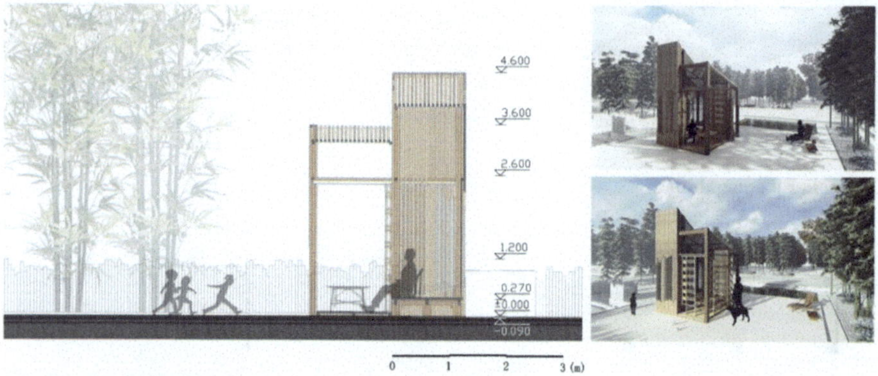

Fig. 3 Space experience and analysis of human activities

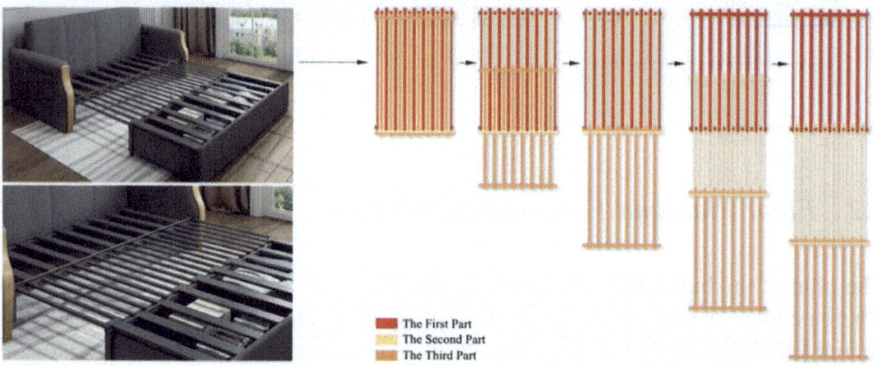

Fig. 4 Inspiration source from furniture. (Left pictures from: https://detail.1688.com/pic/622689974269.html?spm=a261y.8881078.0.0.5d8c6ce2LgLUDx)

The framework is divided into three parts. The first part is installed between the pillars as a fixed part, while the rest has pulleys installed at the grounding points, which ensures that they can be pulled and slid.

The doors on the north side of the building can be pulled out and rotated inward, while the doors on the south side can only be pushed or pulled. By changing the position of the doors, the spatial interface changes between solid and empty, and the privacy level of the space also changes (Fig. 5).

2.4 Sight System Analysis

We analyzed two different states of plans based on the spatial syntax. We converted the virtual reality of the bamboo facade corresponding to the height of human vision

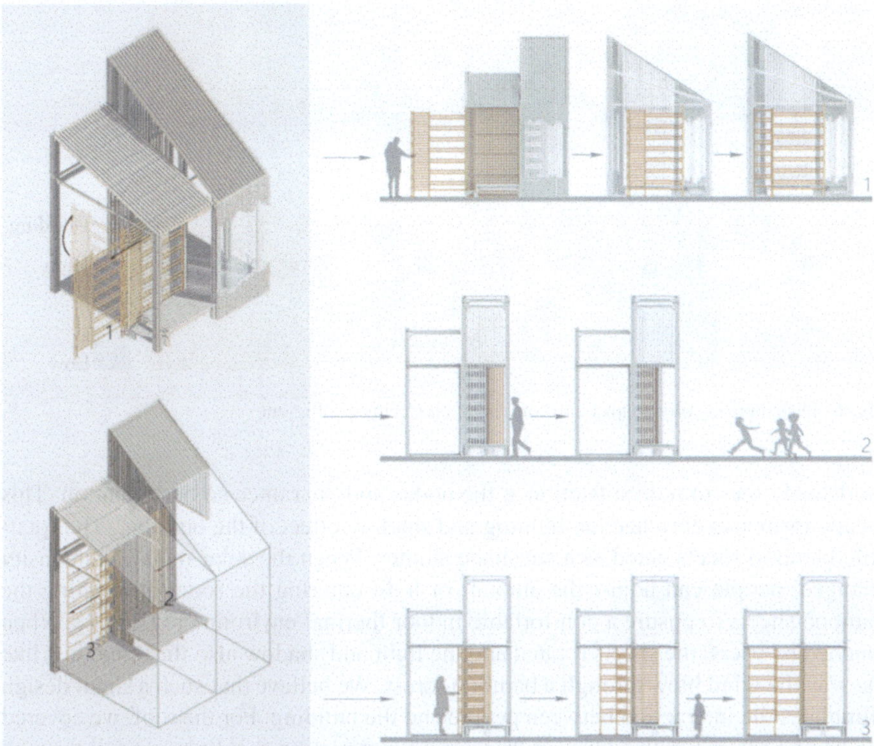

Fig. 5 Analysis of push–pull sliding components

to the building plans and generated distribution color maps based on visual clustering coefficient [6]. We found the results are quite different between the two states, with the door completely open or the door close. Therefore, people can create their ideal space for comfortable by pushing or pulling those components, which is consistent with our original idea to create diverse experiences by promoting interactions between people and buildings (Fig. 6).

2.5 *Skin Details: Elevation and Roof*

In order to enhance the interaction among building, nature, and human, we made a more in-depth detailed design for the facade and roof skin based on the existing form and frame. For the facades, we placed thin, rotatable bamboo sheets in bamboo tubes of different lengths to form a rhythmic decorative facade. The facade texture formed by bamboo slices was different from which formed by the round bamboos, which added a variety of textures to the building, and also showed the variable forms of bamboo as a building material. In order to increase the feasibility of this idea, we

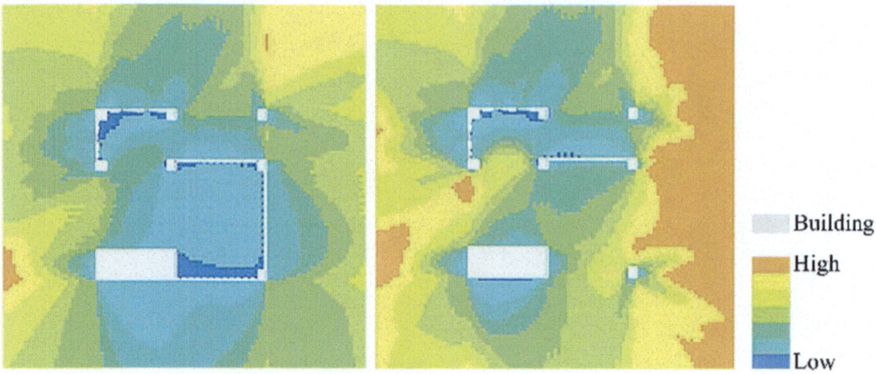

Fig. 6 Distribution color maps based on visual clustering coefficient

used steel pipes to reduce friction at the nodes and guarantee smooth rotation. This facade form also enriched the lighting and shading effect of the building. The rotatable bamboo sheets acted as a sunshade shutter. When the solar radiation intensity changed, people can adjust the amount of light entering the room by rotating the bamboo sheets to ensure a comfortable indoor thermal environment (Fig. 7). When touched by users, the sheets rotated and the light and shadow also flickered, just like the way the wind blew through a bamboo forest. We believe that such a small design stimulates the interaction between people and the building. For the roof, we covered the bamboo frame with a layer of translucent material, so that light can fall through the intervals and make the light and shade effect more magnificent, while at the same time the translucent material can play a role in preventing rains (Fig. 8).

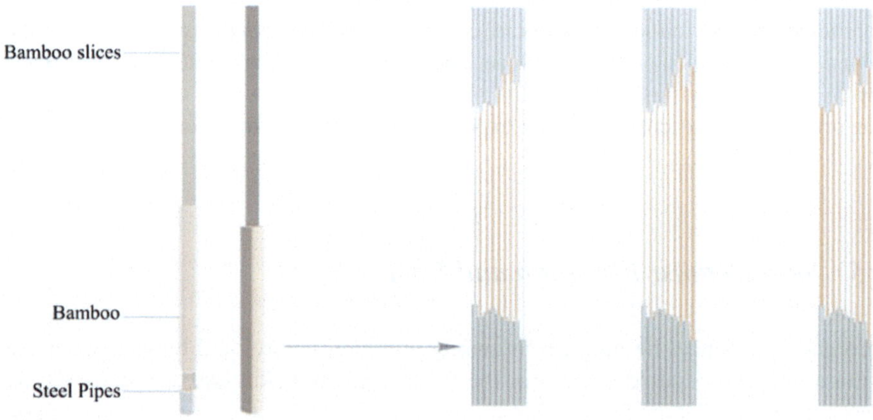

Fig. 7 Rotating the bamboo sheets

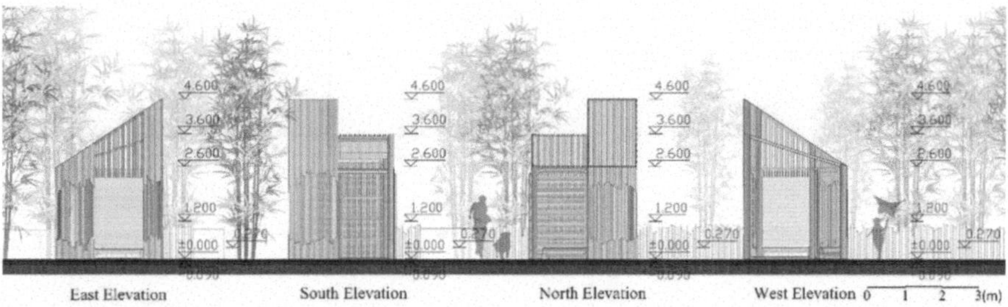

Fig. 8 Elevations

2.6 *Structure Analysis*

The Connection Method of Universal Wheel. When designing the pulley, mobile stability and steering flexibility should be considered. The reasonable structure of the universal wheel largely depends on the stability and the flexibility of the sliding door when it is pushed and pulled. We chose two common connection forms for comparative analysis. One is the shaft-type wheel and the other is the ball-bearing type. By referring to the horizontal speed change and vertical speed change of these two different connecting universal wheels under experimental simulation, it is found that the slope of the speed change curve of the shaft-type wheel is less than which of the ball-bearing type wheel. Under the same level of the horizontal driving force, the ball-bearing type wheel has greater horizontal acceleration and better flexibility, which means under driving force with the same speed, the ball-bearing type wheel is more labor-saving. On a flat road, the speed change of the two types of universal wheels is almost zero, when the universal wheel travels on obstacle roads, the shaft-type universal wheel has a buffer peak while the ball-bearing type wheel travels at a faster speed, showing a cliff-like landing during the landing process without buffering wave crests, which indicates that the axle-type universal wheels are more stable [7].

In order to make the movement of the sliding doors smoother, we lay the guide rails to meet the requirements of a certain level of ground flatness and to meet the smooth performance requirements of the universal wheels. Considering the flexibility of the sliding door movement, we chose the ball-bearing type wheel.

Force analysis of universal wheel. The universal wheel we chose is with a wheel diameter of 100 mm, a wheel width of 5 mm, and a load-bearing capacity of 270 kg.

The density of bamboo was 800 kg/m^3 and the weight of the sliding door was about 3528 N. The relative position between the center of gravity and the universal wheel is shown in the figure as below (Fig. 9).

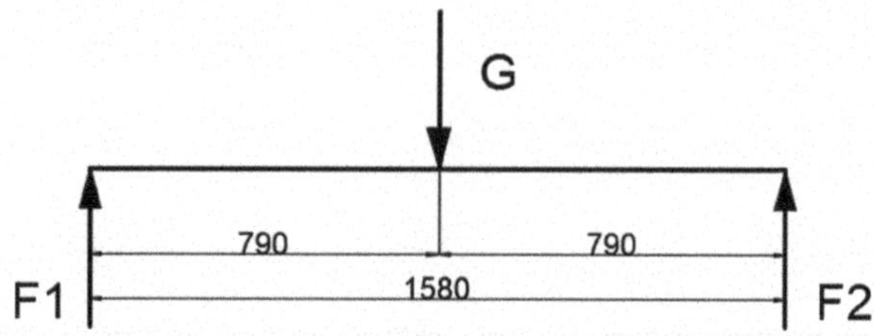

Fig. 9 Schematic diagram of universal wheel forces

According to the vertical force balance equation:

$$\Sigma F_y = 0 \tag{2.1}$$

$$G - F_1 - F_2 = 0 \tag{2.2}$$

The vertical forces on the universal wheel 1 and wheel 2 are:

$$F_1 = F_2 = G/2 = 1,764N \tag{2.3}$$

Design value of load-carrying capacity of universal wheel is:

$$F_{\max} = mg = 270 \times 9.8 = 2,646N \tag{2.4}$$

$$F_1 = 1764N < F_{\max} = 2,646N \tag{2.5}$$

So the selected universal wheel model meets the force requirements.

Stability analysis. The sliding door is moved by the rolling of universal wheels. When the thrust is greater than or equal to the maximum rolling friction, the sliding door moves forward [8]. At this time, the maximum rolling friction couple is:

$$M_{\max} = d \times F_N \tag{2.6}$$

In the formula: d is the rolling friction coefficient. The upper limit of the rolling friction coefficient of the universal wheel and the concrete pavement is 10 mm; F_N is the positive pressure (acting vertically on the supporting surface) of the roller and the supporting surface, which is the weight of the sliding door in the design.

Therefore,

$$M_{\max} = d \times F_N = 10 \times 1.764 = 17.640N \times \text{mm} \tag{2.7}$$

The minimum horizontal thrust F_T required to make the wheel roll can be obtained by the equation:

$$\Sigma M_A(F) = 0 \tag{2.8}$$

$$F_T = M_{\max}/R = 17.640/50 = 352.8N \tag{2.9}$$

Overturning moment produced by horizontal thrust is:

$$M_1 = F_T \times H = 352.8 \times 1.500 = 5.29 \times 10^5 N \times \text{mm} \tag{2.10}$$

Anti-overturning moment of gravity on the sliding door is:

$$M_2 = G \times L/2 = 3.528 \times 1.580/2 = 2.79 \times 10^6 N \times \text{mm} \tag{2.11}$$

As $M_1 < M_2$, so it meets the stability requirements of the mobile process.

2.7 Construction Process and Joints

After the form, function, interactive device, and decorative details were determined, we envisioned the construction process of the building, which provides a clear picture of the design. Once those main components such as foundations, structural columns, beams, and body frames have been constructed, lay the floor and the roof, then install the fixed walls [9–11]. After that, mount the push-and-pull components in the reserved locations [12]. Finally, add the decorative details to complete the whole (Fig. 10).

In order to ensure the feasibility of construction and push–pull components, we consulted a large amount of information about bamboo construction joints, including rope strapping, metal bolts, mortise and tenon joint, and other commonly used construction methods. For push–pull components, we compared various forms of sliding doors and joint connections. We finally decided to use the rail and wheels to reduce friction to ensure that people can move them more easily. One of the wheels is a one-way pulley and the other one is a universal wheel used to cooperate with the rotatable door. The decorative bamboo sheets and tubes of the facades are fixed by round steel pipe to ensure that they are rotatable (Fig. 11).

From form to plan function, from the design of push–pull components to joints connections, from skin details to human activities, we constantly looked up papers and documents, compared varieties of forms to choose the best one, stove to make that every aspect has been detailed considered, to ensure that the original design concept "creating an interesting connection between architecture and its users" is truly realized.

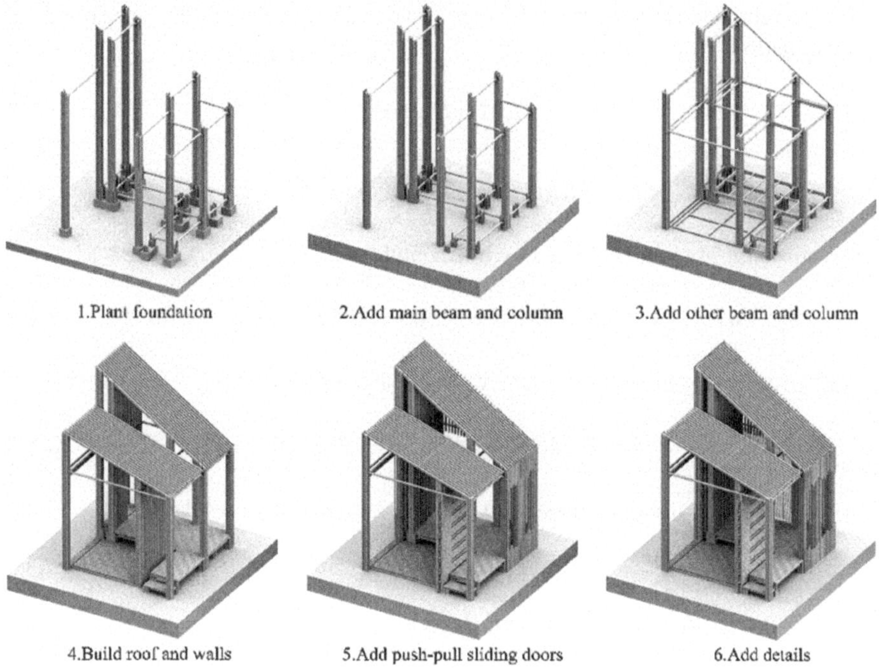

Fig. 10 Construction process

3 Experience

This competition benefited us sufficiently, the team worked closely and made breakthroughs and innovations, our teacher gave us patient guidance and the jury provide us good advice.

As the most important phase of our design, the push–pull sliding components naturally become the focus of the jury. Though the idea is intriguing, it still has shortcomings and there is room for further improvement. The push–pull sliding components work by pure friction, which is simple and can be easily damaged, while there are other forms of force, we can try to improve it. Additionally, we only applied the device to the door and furniture. The door was made as a variable wall that changes the whole space of the building, and the furniture works as a drawer, was made as part of the building. But we didn't expect other parts such as the roof to be variable as well, while a brave attempt just might make the whole design more interesting.

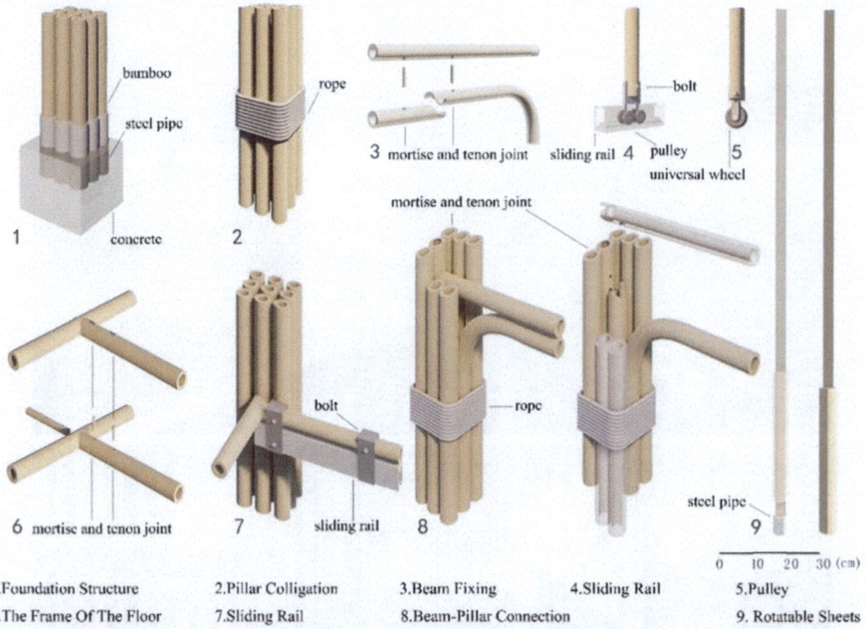

Fig. 11 Joints analysis

4 Renderings

Several renderings of the structure are provided. Figure 2 shows the transparent interface of the building. We designed the blinds, bamboo artifacts that can rotate to show different possibilities of the bamboo material. Figure 3 shows push–pull artifacts formation of the "open" and "close", two different space dialogue with each other, mutual penetration (Fig. 12).

Fig. 12 Renderings of the building

Acknowledgements We gratefully acknowledge the valuable cooperation of Mr. Lin (Faculty adviser), Wei Shi (Master student) and Hao Qiu (Master student) in preparing this article.

References

1. Zhang T. 2016. *Research on construction methods of kinetic facade*. Beijing University of Engineering and Architecture. (in Chinese).
2. Chen Y. 2006. *The study of exhibition space in contemporary Era*. Tongji University (in Chinese).
3. Xiu, D.X. 2013. Research on the interactive mode of interactive installation art. *Art Science and Technology* 26 (06): 241 (in Chinese).
4. Chen, Y. 2006. Installation architecture. *Shanxi Architecture* 15: 31–32 (in Chinese).
5. Yao, C.L., and J. Xu. 2019. Analysis on folding mechanism requirements of furniture standard. *Evaluation and Analysis on Standard* 07: 100–102 (in Chinese).
6. Dalton, R.C., and Q. Dou. 2005. Space syntax and spatial cognition. *World Architecture* 11: 33–37 (in Chinese).
7. Yang, B.J., X.G. Hu, Y.H. Zhong, T.N. Yang, and Z.H. Huang. 2020. Vibration analysis of flat bottom universal wheel. *Mechanical and Electrical Engineering Technology* 02: 77–79 (in Chinese).
8. Wang, C., and B.L. Zhang. 2020. Application of universal wheel in the process of mounting and dismounting Chinese pod form work. *Guangdong Civil Engineering and Construction* 05: 88–91 (in Chinese).
9. Zhang, N., W. F., Bai. 2008. Analysis and improvement of node structure of original bamboo building. *Science Technology and Engineering* 18: 5318–5322+5326 (in Chinese).
10. Li, W.J., and J. Li. 2008. Recycle way of bamboo building. *Huazhong Architecture* 10: 8–10 (in Chinese).
11. Liu, Y.B., and J. Li. 2009. Colombia architect Simon Velez and his bamboo buildings. *World Architecture* 06: 94–97 (in Chinese).
12. Dan, N. 2008. Sliding door: The "Jack of all trades" of space. *Green China* 08: 32–33 (in Chinese).

Bending Blossom

Zhao Zhang, Henry Pantin, and Thomas Wortmann

Abstract Bending Blossom provides visitors with the feeling of being in a bamboo forest. The structure takes advantage of bamboo's elasticity. Bent bamboo poles of the structure transfer horizontal loads and reduce the risk of buckling under vertical loads. A green roof serves as a thermal buffer that provides a comfortable microclimate with natural ventilation. The bending green roof and bamboo create an elegant form in harmony with nature.

Keywords Bamboo · Bending-active · Gridshell · Sustainable design

1 Background

Due to the speed of its growth and material strength, bamboo culms are widely used in Asian vernacular architecture. But in contemporary architecture they are mostly replaced with concrete and steel. One reason for this replacement is the wide variation in bamboo culms' dimensions and structural properties that are caused by different growing environments. But the unique cellular structure of bamboo culms gives them strength not only in compression and tension but especially when the bamboo culms are bent [1].

The ZCB Bamboo Pavilion is a long-span, bending-active bamboo gridshell that is four floors tall (Fig. 1). A team of researchers from CHUK School of Architecture led by Prof. Kristof Crolla designed the pavilion and, in October 2015, completed

Z. Zhang (✉) · H. Pantin · T. Wortmann
Department of Architecture and Design, Xi'an Jiaotong-Liverpool University, Suzhou, China

H. Pantin
e-mail: h.pantin18@student.xjtlu.edu.cn

T. Wortmann
e-mail: thomas.wortmann@icd.uni-stuttgart.de

T. Wortmann
Chair for Computing in Architecture, Institute for Computational Design and Construction, Faculty for Architecture and Urban Planning, University of Stuttgart, Stuttgart, Germany

K. Liu et al. (eds.), *2019 International Bamboo Construction Competition*, Springer Tracts in Civil Engineering, https://doi.org/10.1007/978-3-030-91990-0_14

Fig. 1 ZCB bamboo pavilion [3]

it in Kowloon Bay, Hong Kong, China. Three layers of bamboo culms, 475 in total, were bent and hand-tied into a bending-active, triangulated diagrid onsite.

The pavilion's structure was developed and tested with a series of digital and physical models, including bamboo prototypes [2]. The researchers "flattened" the digital model into straight lines to transfer the culms' intersection points from the digital model onto the physical bamboo culms. The bamboo culms were hand-tied with metal wires because traditional knots are not fireproof [3].

As the ZCB Pavilion demonstrates, active-bending bamboo structures utilize bamboo's natural properties better than pre-bent, curved bamboo structures. But, in practice, there are many more examples of the latter than the former, probably due to the difficulty of designing active-bending structures, especially with an anisotropic, natural material. In other words, the potential of bamboo for efficient and sustainable, active-bending structures still needs to be explored and exploited more fully [4]. For this reason, we designed our pavilion as an active bending structure.

Fig. 2 Concept design sketches

2 Concept Design

The pavilion provides a raised platform to sit and relax on, and from which to observe passers-by. An eye-catching, cantilevered canopy offers shade and rain protection for this platform. The curved shape of the canopy is inspired by an inverted blossom and topped with a green roof to more harmoniously blend into the surrounding horticultural expo (Fig. 2).

3 Design Development

The pavilion consists almost entirely of bamboo. A bolted frame of primary bamboo culms (diameter 10 cm) is filled in with a secondary grid of smaller, tied bamboo culms (diameter 5 cm). The sides of the pavilion are covered with bamboo weave and the raised floor with bamboo laminate (thickness 3 cm). The pavilion's base fits in 3×3 m^2 (Fig. 3b), and the canopy in 4×4 m^2 (Fig. 3a). The pavilion's ceiling is covered with a green roof (Figs. 4 and 5). This green roof helps to maintain the pavilion's shape by weighing down the bending-active structure.

The pavilion is a lightweight, bending-active structure, whose shape and stiffness are created through bending the primary bamboo culms. Bächer et al. [5] state that the bending-active, large-span shell structures of the Mannheim Multihalle were made possible by prefabricated regular grids without shear rigidity that were bent and braced on site. Our pavilion follows a similar approach: the primary bamboo culms are bent by connecting them and loading them with the green roof. We have demonstrated the feasibility of this structural concept with a physical 1:10 prototype (Fig. 6). The secondary culms connect the primary culms into a grid structure.

There are three principles to prevent bamboo from decaying: (1) Shielding from direct sunlight or ultraviolet rays, (2) avoiding exposure to rain and water, and (3) prophylactic treatment to protect from biotic attack [6]. Accordingly, our bamboo structure is covered with a bamboo weave to protect it from UV and humidity.

The pavilion is naturally ventilated, with additional cooling provided by the green roof. The green roof is to be planted with flowers according to the context of the

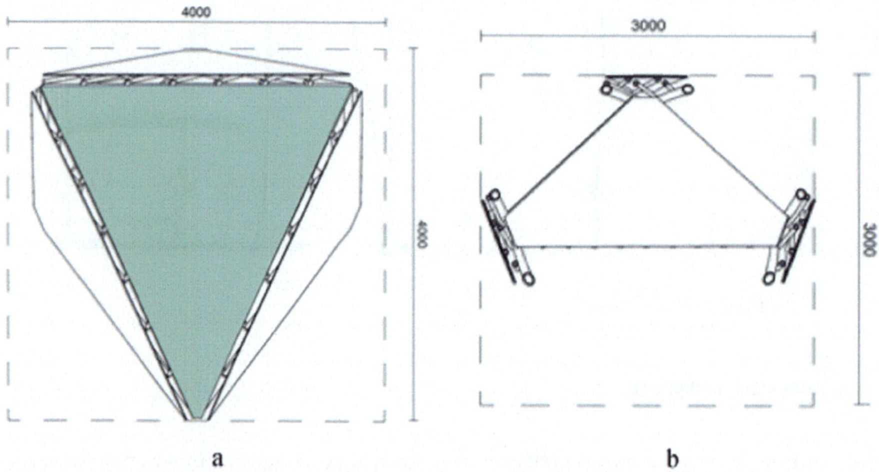

Fig. 3 Roof plan (**a**); Floor plan (**b**)

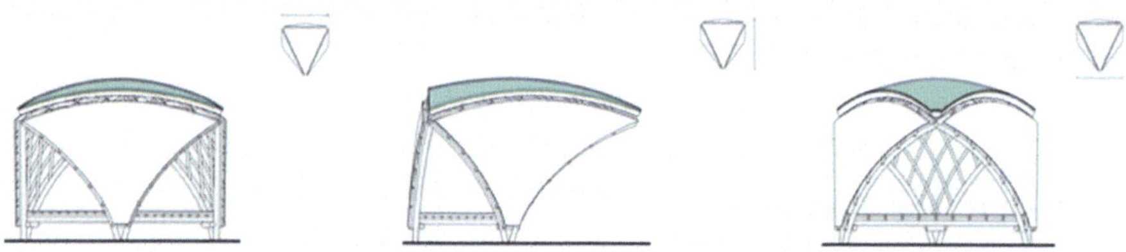

Fig. 4 Elevations

Fig. 5 Section

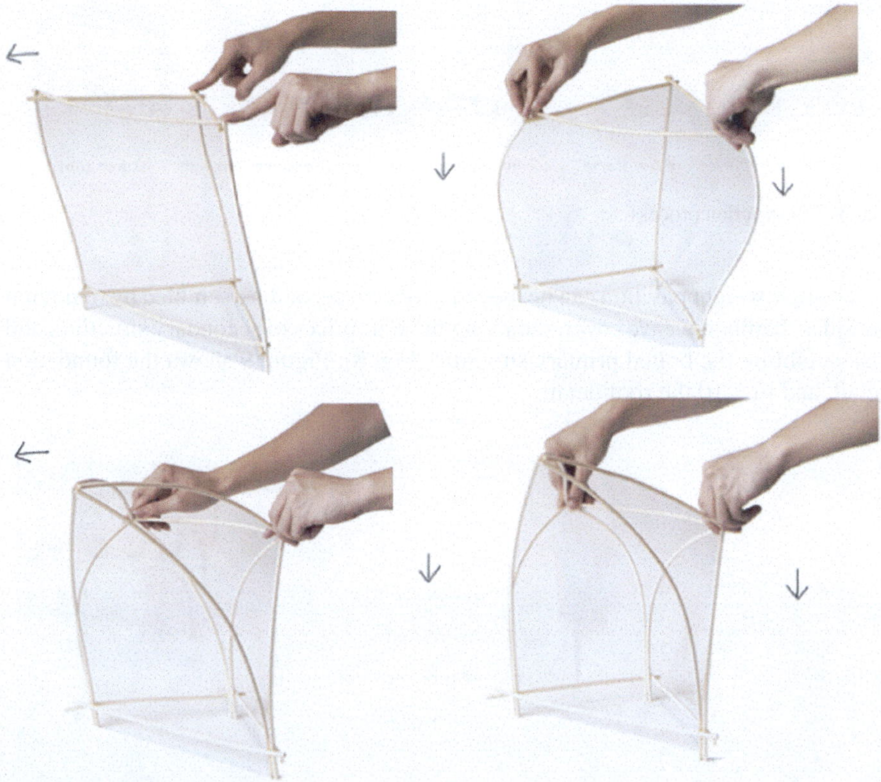

Fig. 6 Physical model testing

horticultural expo. The canopy's curved shape assists in the drainage of rainwater. The green roof is watered mainly by rainwater. When the green roof needs watering, there is a water pump and hose, integrated into the bamboo culms, to water the grass (Fig. 7).

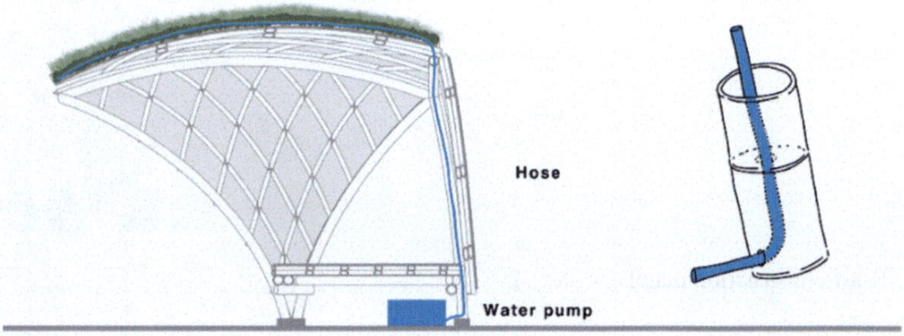

Fig. 7 Irrigation system

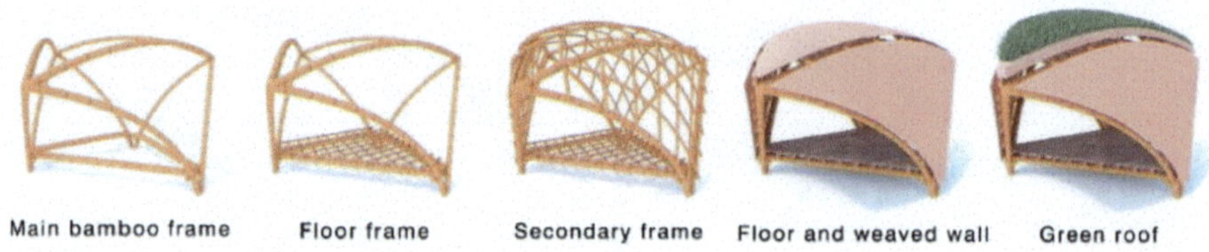

Fig. 8 Construction process

The lightweight pavilion can be moved in one piece, or disassembled by removing the sides' bamboo weave cover, removing the four prices of secondary structure, and disassembling the bolted primary structure (Fig. 8). Figure 9 shows the foundation detail, and Fig. 10 the roof detail.

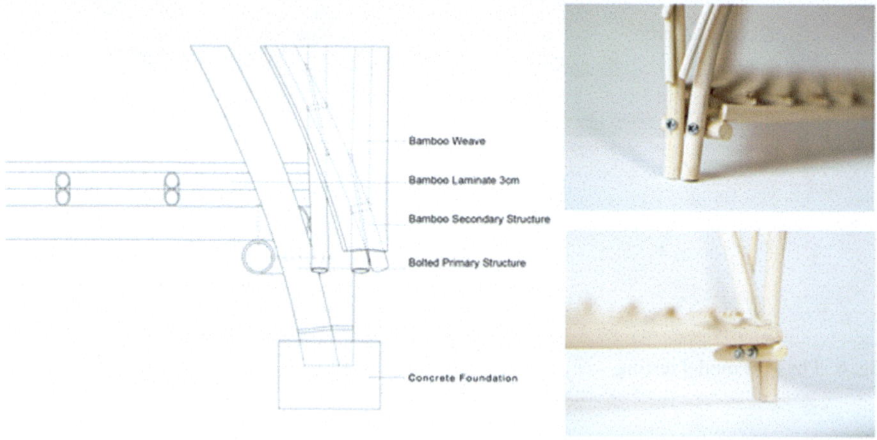

Fig. 9 Foundation construction detail

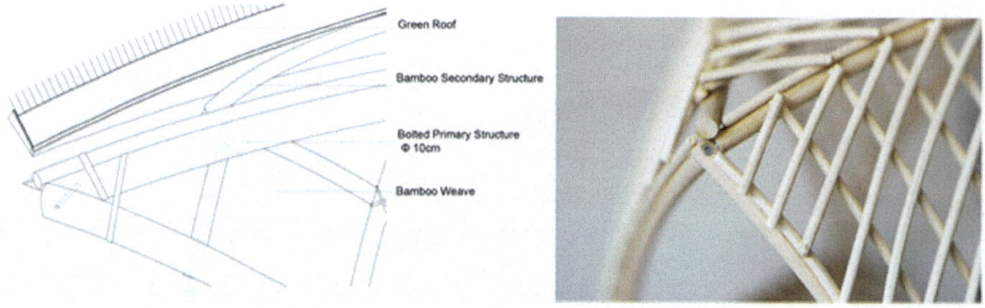

Fig. 10 Roof construction detail

4 Renderings

The renderings in Figs. 11 and 12 show that the bamboo pavilion is a bending-active gridshell. The bolted frame of primary bamboo culms is filled with a grid of smaller, secondary bamboo culms. The structure is covered by bamboo weave to provide shading to visitors. One of the structure's corners is extended, to shelter visitors when they are sitting in the pavilion or standing under the roof (Fig. 12). The bamboo pavilion is suitable to be located in urban and natural environments.

Fig. 11 Rendering of outside view

Fig. 12 Rendering of inside view

5 Conclusions

Our tests with a physical model show that a non-bending structure is not stiff enough and needs additional bracing when dealing with horizontal forces. But the active-bending structure shows a strong ability to prevent buckling and bending from vertical and horizontal forces. Bamboo is good at bending, which provides great potential to create a curved, structural form.

This pavilion design demonstrates that bamboo has multiple advantages as a sustainable building material. Ecologically, for sequestering carbon and providing natural habitat, aesthetically, as a natural material, structurally, due to its strength in active bending, and for construction, since it is easy to build with.

References

1. Crolla, K. 2017. Building indeterminacy modelling—the "ZCB Bamboo Pavilion" as a case study on nonstandard construction from natural materials. *Visualization in Engineering* 5: 15.
2. Crolla, K. 2018. Bending bamboo rules: Beyond century-old typologies. *Journal of Architectural Education* 72 (1): 135–145.
3. Crolla, K., and A. Fingrut. 2016. The design and construction of a bending-active gridshell from natural bamboo.
4. Rockwood, D. 2015. *Bamboo gridshells*. London: Routledge.
5. Bächer, M., B. Burkhardt, and F. Otto. 1978. IL13—Mannheim Multihalle. Stuttgart: Institute for Lightweight Structures, University of Stuttgart.
6. López, O.H. 2003. Bamboo: the gift of the gods. D'VINNI LTDA.

Ternary Shed: An Exemplar Design for Bamboo Construction

Harry F. Mills, Mateo Gutierrez Gonzalez, and Hamish Buchhorn

Abstract The Ternary Shed is an exemplar bamboo building design that demonstrates bamboo's mechanical and aesthetical characteristics in construction. The design draws on a triad of knowledge from Latin American low-cost construction techniques, Australian timber construction methods, and contemporary Chinese bamboo design. The Ternary Shed is a two-story pavilion made of round bamboo poles and engineered bamboo products, entirely designed for prefabrication to enable rapid, accessible on-site construction. The flexibility and adaptable layout results in a structure that can be deployed for many uses and typologies worldwide. The design was submitted for the International Bamboo Construction Competition 2019 and presented at the 2019 International Horticultural Exposition in Beijing, China, where it placed third overall. The experience of participating in the competition and the design process was an opportunity to raise public awareness about bamboo's potential use in construction and to demonstrate its versatility and strength as a renewable building material.

Keywords Round bamboo · Laminated bamboo · Renewable building products · Resource conscious design · Construction

1 Introduction

The Ternary Shed is a two-story bamboo pavilion designed to promote bamboo construction and address the needs and challenges presented in the International Bamboo Construction Competition 2019 (IBCC 2019). The project raises public awareness about bamboo's opportunity and potential for use in construction. The

H. F. Mills (✉) · M. G. Gonzalez
School of Civil Engineering, University of Queensland, Brisbane, Australia
e-mail: harry.mills@uqconnect.edu.au

M. G. Gonzalez
Cusp Building Solutions, Tasmania, Australia

H. Buchhorn
School of Architecture, University of Queensland, Brisbane, Australia

K. Liu et al. (eds.), *2019 International Bamboo Construction Competition*, Springer Tracts in Civil Engineering, https://doi.org/10.1007/978-3-030-91990-0_15

Ternary Shed has been designed to be an adaptable pavilion that can function as a temporary medical facility, warehouse, office building, community educational hub, campground shelter, tea house, bar, cafe, restaurant, and other uses. The design combines the traditional round bamboo design techniques from Latin-America [1], the contemporary concepts of timber design applied in Australia, and modern engineered bamboo materials created in China [2]. Most of the Shed's building components can be prefabricated offsite and transported for on-site construction. Doing so enables rapid construction, a small workforce, lower labour cost, and safe construction practices. After use, the building can be deconstructed and reused in other future applications.

The structure is designed to withstand extreme environmental events such as earthquakes, wind events, snow, rain, and typical gravitational dead and live loads. *Phyllostachys pubescens* (Moso bamboo) is the primary load-bearing material. Round bamboo poles provide vertical and lateral stability creating a 3D truss to enable the material to use its remarkable axial strength to withstand tensile and compressive axial forces. In contrast, engineered bamboo members on the second floor efficiently take the vertical loads' bending moments and shear forces. Furthermore, the laminated bamboo beams and flooring deliver a finished product of excellent visual appearance. The roof is a curved grid that transfers the loads to the prominent vertical members. The primary structural support for the roof is a central composite beam that cantilevers over the opening area of the pavilion. By designing the cantilever over the entrance, the roof creates a sense of welcoming to the space whilst highlighting the capabilities of bamboo, as demonstrated in Fig. 1.

2 Concept Design

The design of the Ternary Shed considers the need and opportunity to showcase bamboo's ability to be affordable and prefabricated in the short span of 24 h. The concept draws on the state-of-the-art residential, community, and temporary structures, with precedent projects listed in Table 1. The design highlights bamboo's structural capacity through a round and laminated system to support a two-story building with three anchoring supports on the ground plane. The structure is exposed to demonstrate the ability of round bamboo and bamboo veneers. The bamboo systems work in conjunction with engineered timber for the stairwell structure. Semi-transparent polycarbonate panels in two neutral pastel tones of green and white form the cladding of the roof. Areas typically requiring a balustrade are designed with a recycled textile fishnet, seen in the stairwell and second floor. The net allows for ventilation, natural light, continuity of eyesight and keeps the attention pinned to the bamboo structure. Different levels of bamboo construction technology inform the overall architectonic and structural design: (i) low-tech joins like lashing connections with steel wire, (ii) mid-tech links on the joints of round bamboo, (iii) using steel bolts, dowels and mortar refill, and (iv) high-tech connections using flexible and adaptable steel rods and plates. Several mixed construction techniques are utilised along the prefabrication

Fig. 1 Ternary Shed, perspective of campsite facility

and construction process. The structure enables the pavilion to remain full ventilated, capturing summer breezes while providing extensive shading to occupants.

For low-tech construction detailing, Latin-American construction methods such as round bamboo combined with bolted connections filled with mortar are used to connect round bamboo to round bamboo members [7]. The roof members also use composite bamboo beams joined by bolts and metal bands [8]. More sophisticated construction technologies such as laminated bamboo bearers and joists and laminated bamboo flooring are incorporated on the secondary level [9]. Other bamboo materials such as bamboo scrimber (Strand woven bamboo) and bamboo veneer [2] are proposed for non-structural wall partitions, cabinetry and furniture. These systems and preliminary sketches are displayed in Fig. 2.

The structure can be assembled rapidly due to the prefabrication of beams and round bamboo members. The roof and facades are designed to be prefabricated offsite with Computer Numerical Control (CNC) routers. The laminated bamboo veneer wall, partition, and stair elements are modularised for accessible construction and deconstruction. With the modularity of the design, the Ternary Shed layout can be altered to multiple variations depending on the context and occupier needs. For example, four layouts were proposed for the IBCC without impacting the structural system and load-bearing elements. Layout changes occur with the shifting of cabinetry, partition walls and other non-structural members. The following section highlights the different layout options of the Ternary Shed.

Table 1 Precedent projects

Project image				
Project title	Pachacamac [3]	RAUM pavilion [4]	Tower house [5]	Comedor uumbal [6]
Designer	Studio Tom Emerson ETH Zurich, Taller 5 PC University of Peru	Overtreders W	Austin Maynard architects	Lucila Aguilar Arquitectos
Construction	2019	2019	2014	2016
Features	Woven white polyester textile, timber & frames bamboo	Polycarbonate and wooden frames with exposed structure	Multifunctional uses. Vertical space connected/protected with a net	Cantilever roof on round bamboo, round bamboo grid shell

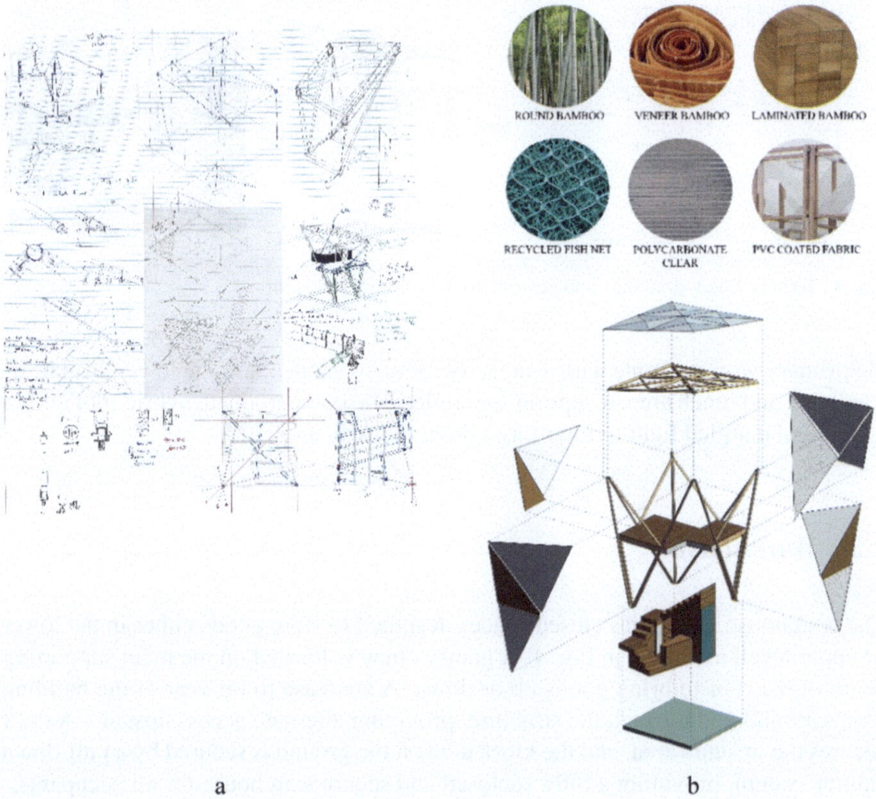

Fig. 2 Process sketches **a**; and materiality and exploded isometric view **b**

2.1 Office

The office layout presented in Fig. 3 is a semi-enclosed space that provides a well-ventilated, natural light-filled area with ample privacy and security. Exposed bamboo and timber panels give comfort and enhance work productivity. The upper level is

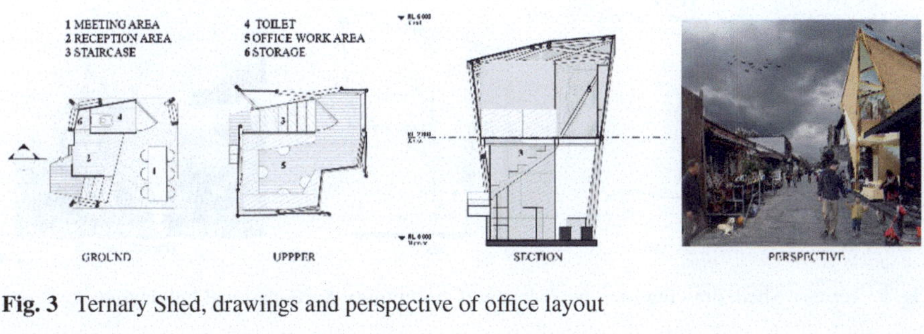

Fig. 3 Ternary Shed, drawings and perspective of office layout

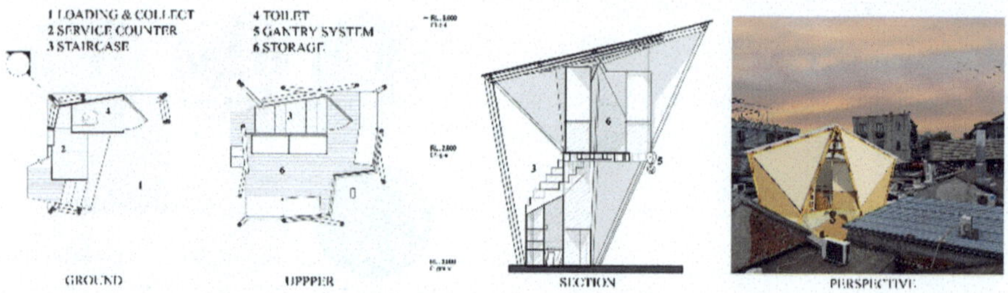

Fig. 4 Ternary Shed, drawings and perspective of warehouse layout

the primary working zone with a capacity for up to six people. The lower level hosts a meeting and lunch area, reception, and toilet. The polycarbonate hybrid roof allows for natural dappled light to filter throughout the structure.

2.2 *Warehouse*

The warehouse layout has closed spaces designed to store goods either in the lower or upper level, as shown in Fig. 4. A gantry crane is located on the main supporting beam of the roof to bring goods up or down. A staircase to the rear of the building wraps around and through the structure, providing alternate access upstairs. A door secures the upstairs area, and the kiosk area on the ground is secured by a pull-down shutter system, providing a fully enclosed and secure warehouse for all occupants.

2.3 *Community and Educational Hub*

The community and educational hub displayed in Fig. 5 is a multifunctional space that hosts several social activities. The lower level contains an informal gathering zone with an interactive screen adjacent to a service counter and storage station. An

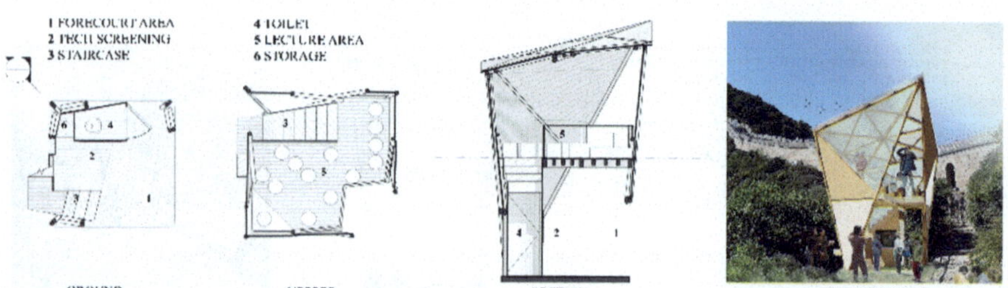

Fig. 5 Ternary Shed, drawings and perspective of community & educational hub layout

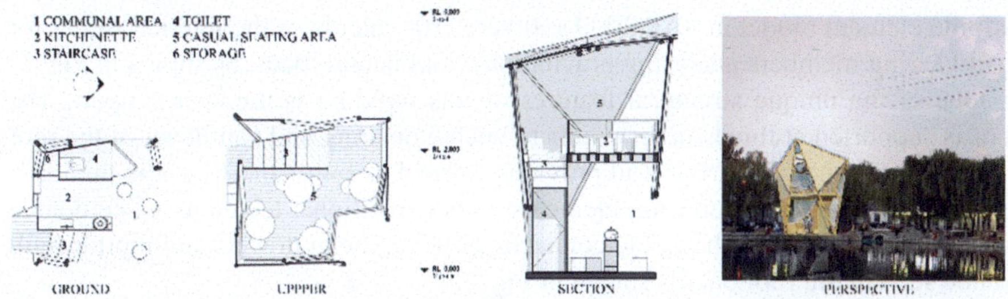

Fig. 6 Ternary Shed, drawings and perspective of and relaxation house layout

enclosed private pod is positioned at the perimeter of the gathering zone and storage area. The private pod can be used as a self-contained toilet and shower or a temporary sleeping shelter. The upper level has a gathering zone with a raised platform for presentations and speaker events. The area is equipped with mobile furniture that can be assembled in various layouts for lectures, talks, seminars, workshops and other community or educational activities.

2.4 *Tea and Relaxation House*

The tea and relaxation house presented in Fig. 6 is an open layout for social spaces such as a bar, cafe, restaurant, or communal kitchen. A small kitchenette and prep area with a few seats are located at the lower level. The upper level contains multiple tables that can seat up to twelve people. The room is well lit and open to attract natural ventilation while also providing shelter for occupants on both levels. The area is designed to be accessed 24/7 and provides sanitary and cooking facilities.

3 Design Development

3.1 *Structural Background*

Bamboo is the primary load-bearing material for both the floor slabs and supporting columns of the Ternary Shed. Round bamboo members are used when subjected to axial forces parallel to the fibres. Bamboo columns and bracings act as a 3D truss to provide lateral stability. The 3D truss has been designed to withstand wind and seismic forces, transmitting all gravitational loads from the secondary level and roof to the ground plane. Laminated bamboo beams work in conjunction to provide the secondary level's required shear and bending capacity and restrain lateral buckling of the round bamboo elements under compression to increase the compressive capacity.

A finite element model in SAP 2000 software [10] calculates the internal forces for load-bearing members analysing gravitational and lateral loads, as shown in Fig. 7.

One of the unique structural features of this building is the roof support. The roof is supported at three corner points to enable opening and cantilever of the roof above the entrance. The roof grid provides vertical stability and transfers the loads to the main round bamboo truss elements. Other traditional techniques like double bamboo beams and bamboo pole connections using metal dowels and mortar refill can be seen in the roof details shown in Fig. 8.

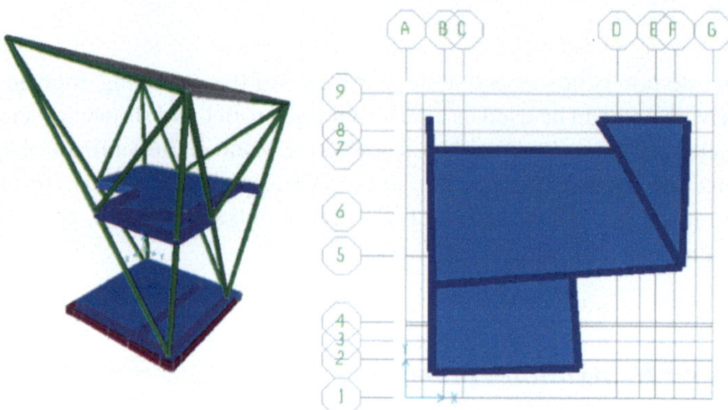

Fig. 7 3D model of the structure and plan view from the upper-level slab

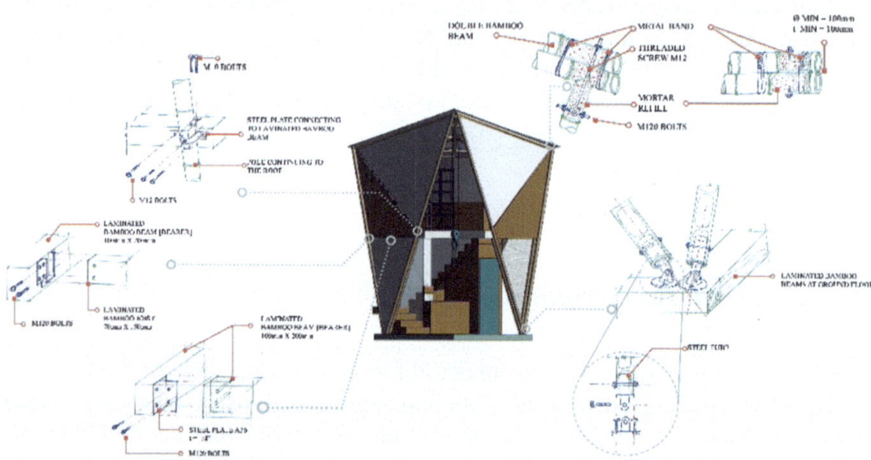

Fig. 8 3D model with construction details

3.2 Design Actions

Permanent and imposed loads: Permanent and imposed loads were considered following the requirements of Australian Standard AS 1170 [11]. Lower and upper-level slabs were designed to support an imposed load of 3 kN/m^2, recommended for offices, workrooms, and public areas. The roof was also designed to support vertical loads required to conduct repairs or provide access to that area, considering an imposed load of 1.5 kN/m^2. A detailed gravitational dead load assessment can be seen in Table 2.

Seismic and wind loads: According to the Chinese code of seismic design of buildings [12], Beijing has been classified with an earthquake-resistant fortification intensity of 7, and the basic design acceleration is 0.20 g. To account for the worst-case scenario, a basic design acceleration of 0.90 g was assumed. A standard value of wind load pressure of 0.45 kN/m^2 was considered for the wind actions to design the roof and the façade.

Load combinations suggested in the Australian Standard were considered for the design of the structure. Drifts were checked for seismic and wind actions, allowing for the structure not to deform more than 1% of the story height. Due to the high stability provided by the 3D truss, lateral displacements in the structure are considerably low. The structure is very stiff in all directions. Figure 9 shows the assigned permanent gravitational loads and the deformed structure under seismic load in the Y direction.

Structural design for critical members: The allowable stresses method and the characteristic values provided by the Colombian Bamboo Construction Code [13] were considered in the design of the laminated and round bamboo members, respectively. Other design frameworks such as ISO 22156 [14] were cited for all the critical members. The structural analysis indicated that the columns must be composed of at least two bamboo poles of a minimum diameter of 100 mm, and a minimum wall thickness of 10 mm due to the magnitude of compressive and tensile axial forces in the lower level. The main load-bearing beam in the roof should be made of a minimum of two poles of the same size. The bending stresses govern the

Table 2 Dead loads assessment

Permanent load: secondary level slab		
Bamboo flooring (Laminated bamboo)	0.24	kN/m^2
Services and fittings	0.15	kN/m^2
Joists and steel connectors	0.32	kN/m^2
Subtotal permanent load	**0.71**	**kN/m^2**
Permanent load: roof		
Bamboo mat	0.29	kN/m^2
Polycarbonate or transparent	0.20	kN/m^2
Services and fittings	0.15	kN/m^2
Battens and steel connectors	0.15	kN/m^2
Subtotal permanent load	**0.79**	**kN/m^2**

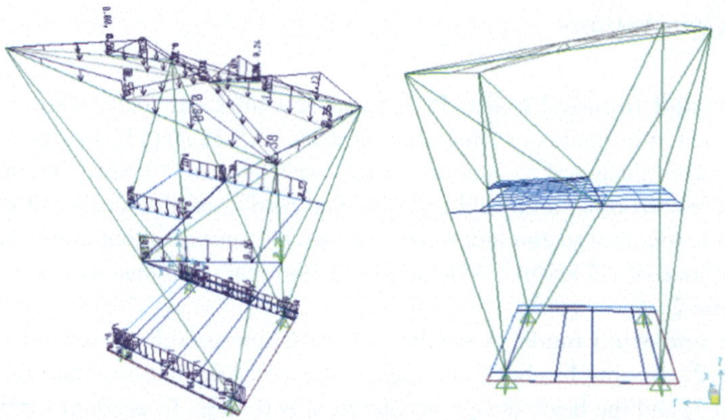

Fig. 9. 3D model with assigned gravitational loads (left) and the deformed structure under seismic load in Y direction (right)

final dimension of the laminated bamboo beams, resulting in rectangular loading bearers of 100 mm × 200 mm, and joists of 70 mm × 150 mm, separated maximum by 500 mm. The connections between the joists and bearers should be made by steel bolts and plates, as shown in the details presented in Fig. 8.

3.3 Construction Process

As represented in the scale 1:200 physical model in Fig. 10, the Ternary Shed explores multiple structural details to highlight the possibilities of bamboo as a renewable building material. The design reveals unique forms, shapes, and spaces across all levels, providing shelter, access to natural light, and ventilation. All design elements, including the laminated bamboo bearers, joists, floor, beams, wall partitions, round bamboo poles, steel connection tubes and plates, and triangular roof structure, are fabricated offsite. The elements can be assembled with low-tech methods on-site with a small number of people. With scaffolding or a scissor lift, the second level and roof elements can be installed quickly.

Fig. 10 Physical model

The round bamboo pole structure is assembled and attached to the ground plane, followed by the lifting and fitting of the second level bearers, joists and floorboards. The roof beams and round Bamboo bracing are attached to the perimeter structure and prepared for the roof elements. The lower-level wall partitions, cabinetry and other non-structural members can be set up internally. The triangular bamboo framed polycarbonate panels can finally be lifted onto and attached to the roof supports. The connections between the roof beams, round bamboo bracings and columns require scaffolding or scissor lifts to reach the apex.

4 Experience at INBAR IBCC 2019 Competition

The Ternary Shed proposal for the IBCC 2019 Competition was a rewarding and enjoyable experience. Visiting and presenting the proposal at the Bamboo INBAR Pavilion at the 2019 International Horticultural Expo was a unique opportunity that the team thoroughly appreciates. We are immensely pleased to have represented Australia in front of an expert panel of industry and academic leaders and be awarded third overall. The transdisciplinary nature of our team brought new collaborative experiences for undergraduate and postgraduate students from Architecture, Engineering and Design.

5 Conclusive Remarks

Bamboo is a low-embodied carbon material, and in comparison to other high-embodied carbon alternatives like concrete and steel, is natural and renewable. Its versatility and robustness mean bamboo can provide enthralling and economical infrastructure solutions when correctly used. With its fast growth rate and high rotation ability, bamboo's use in construction can increase to become an essential building material that can be deployed for uses and typologies worldwide.

The Ternary Shed proposal demonstrates several bamboo construction solutions for creating an adaptable and resource conscious design. The products and systems incorporate traditional low-cost and low-tech methods and contemporary digital fabrication techniques to enable rapid, accessible construction and deconstruction. Our transdisciplinary team presents an exemplar for bamboo construction by drawing on a triad of knowledge from Latin America, Australia, and China.

Acknowledgements The team would like to acknowledge the support from all involved at the Schools of Architecture and Civil Engineering, and the Employability Centre at the University of Queensland. Thank you to INBAR for the opportunity to participate in the competition and for the experience of presenting our proposal at the 2019 International Horticultural Exposition in China. Lastly, thanks to Kim Baber for his guidance and mentorship across the design and competition period, and Carla Amaral for her assistance prototyping and model making.

References

1. Hidalgo López, O. 2003. *Bamboo: The gift of the gods*, Bogotá, Colombia S.A.
2. Sharma, B., A. Gatoo, M. Bock, H. Mulligan, and M. Ramage. 2015. Engineered Bamboo: State of the art. *Proceedings of the Institution of Civil Engineers: Construction Materials* 168: 57–67. https://doi.org/10.1680/coma.14.00020.
3. Studio Tom Emerson ETH Zurich, Facultad de Arquitectura y Urbanismo, PUCP Lima, Pachacámac: A Room for Archaeologists and Kids, Room Archaeol. Kids. (2018). https://emerson.arch.ethz.ch/construction/pachacmac-a-room-for-archaeologists-and-kids.
4. Overtreders, W. 2018. *Raum project*, RAUM. https://en.overtreders-w.nl/raum.
5. Austin Maynard Architects. 2015. Tower house, tower house. https://maynardarchitects.com/865004006353.
6. Lucila Aguilar Arquitectos, La Ceiba, La Ceiba. 2016. https://www.lucilaaguilar.com/la-ceiba/.
7. Correal, J.F. 2016. 14—Bamboo design and construction, in ed. K.A. Harries and B. Sharma, Nonconv Vernac Constr Mater, Woodhead Publishing, pp. 393–431. https://doi.org/10.1016/B978-0-08-100038-0.00014-7.
8. Kaminski, S., A. Lawrence and D. Trujillo. 2019. Structural use of bamboo. Part 1: Introduction to bamboo, The Institution of Structural Engineers,. https://www.istructe.org/journal/volumes/volume-94-(2016)/issue-8/structural-use-of-bamboo-part-1-introduction-to-b/.
9. Sharma, B., A. Gatóo, M. Bock, and M. Ramage. 2015. Engineered bamboo for structural applications. *Construction and Building Materials* 81: 66–73. https://doi.org/10.1016/j.conbuildmat.2015.01.077.
10. CSI, SAP2000 V-14 Integrated finite element analysis and design of structures basic analysis reference manual, computers and structures Inc., Berkeley, CA, United States, 2010.
11. SAI Global. 2002. AS/NZS 1170.0:2002.
12. Ministry of Housing and Urban-rural Development of the People's Republic of China, GB 50011–2010, China Architecture & Building Press, Beijing, PRC, 2010. https://www.chinesestandard.net/PDF/BOOK.aspx/GB50011-2010.
13. AIS, Reglamento Colombiano de Construcción Sismo Resistente NSR-10 Titulo G—Estructuras de Madera y Estructuras de Guadua, in: Asociación Colombiana de Ingeniería Sísmica, Bogota, Colombia, 2010: p. 168.
14. ISO, ISO 22156:2004, ISO, 2004. https://www.iso.org/standard/36149.html.

Bamboo's Verve

Biao Huang, Tao Hu, Zehui Jiang, Xinrui Wei, Jiping Ma, Junfeng Xiong, and Zhen Hu

Abstract Inspired by the biological process of natural growth of bamboo, this design simulates the growth characteristics of bamboo rhizome, and uses the good toughness of bamboo materials at the present stage to form a bamboo landscape architecture suitable for bending, folding, and standing, named "Bamboo's Verve". In line with the theme of "Better City, Better Life" of 2019 Beijing International Horticultural Exposition, the design embodies the green development, fusion and blooming, and colorful environment.

Keywords "Bamboo's Verve" · Bamboo architecture · Structural system · Environmental protection

1 The Conception

1.1 Inspiration Source

Bamboo Gardening: Build a multi-functional garden space by simulating the growth characteristics of bamboo rhizome. The proposed project aims at interpret the evolving space scenes created by different bamboo materials and show the "three realms" of bamboo by traditional Chinese gardening art: physiological characteristics, painting scene and artistic conception (Fig. 1).

What are physiological characteristics? Bamboo rhizome refers to the slender underground stem of bamboo, which runs horizontally. There are nodes on the bamboo rhizome, on which rhizome roots grow. There are also buds on the side of the nodes, some of which develop into shoots and others become new rhizomes.

B. Huang (✉) · T. Hu · Z. Jiang · X. Wei · J. Ma · Z. Hu
International Centre for Bamboo and Rattan, Beijing, China
e-mail: huangbiao@icbr.ac.cn

J. Xiong
Chongqing Jiaotong University, Chongqing, China

Z. Hu
Beijing Forestry University, Beijing, China

K. Liu et al. (eds.), *2019 International Bamboo Construction Competition*, Springer Tracts in Civil Engineering, https://doi.org/10.1007/978-3-030-91990-0_16

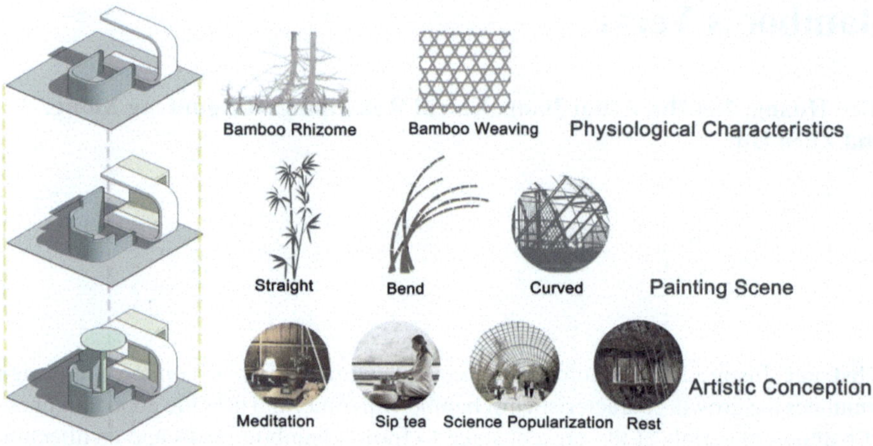

Fig. 1 Inspiration source

In a bamboo forest, bamboo culms above the ground are separated. However, underground bamboo rhizomes are united into one, which originated from one or a few "bamboo trees". In the Chinese Character Dictionary, "weaving" is interpreted as "organizing by using thin strips or strips of things". Bamboo weaving takes bamboo sticks or bamboo strips as materials. Horizontal and longitudinal interweaving is carried out by means of picking and pressing [1].

What is a painting scene? Inspired by the process of bamboo's natural growth, this design imitates the growth characteristics of bamboo rhizome and makes use of the good toughness of bamboo to form a bendable and foldable bamboo building. As an ornamental and functional space, the bamboo building combines its bamboo landscape with the exhibition in INBAR Pavilion to form a deep scroll picture of bamboo (Fig. 2) [2], which can attract different groups of tourists, such as children, artists, and educators.

Fig. 2 Secluded and Restful Bamboo Grove and Stone, an ancient Chinese painting by Gu An

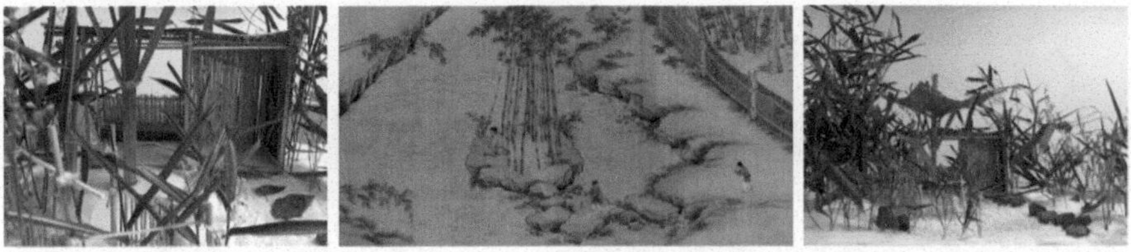

Fig. 3 *Du Paradise*, an ancient Chinese painting by Qiu Ying

What is artistic conception? By exploring the space order in the natural bamboo landscape and the different utilizations of bamboo materials, we provide the space with functions as meditating, educating, rest and sipping tea. It's quiet in motion and dynamic in silence, giving people an artistic conception of being as empty as bamboo, elegant and free from vulgarity. How to expand the traditional Chinese art in bamboo's verve? See Figs. 2 and 3 for details, showing the scenery composition.

1.2 Design Ideas

Inspired by the biological process of the natural growth of bamboo, this design simulates the growth characteristics of bamboo rhizome and uses the good toughness of bamboo to form a bamboo landscape architecture that is "suitable for bending, folding and standing". It not only becomes an ornamental and useful landscape but also meets the needs of all kinds of tourists (children, teenagers, the elders, artists, etc.) in INBAR Garden. Therefore, the name of the bamboo building was given as "Bamboo's Verve" (Fig. 4). "Bamboo's Verve" means "bamboo shows charm and vitality", which is in line with the theme of "Better City, Better Life" of the 2019 Beijing International Horticultural Exposition, and reflects the green development of INBAR Garden.

2 Design Process

2.1 Design Principles

Several design principles are showed as follows: (1) All kinds of materials need to meet engineering design specifications in China; (2) The calculated data of the model meets relevant requirements; (3) Different effective connection modes are designed for different nodes; (4) The design of traffic flow line is reasonable to ensure the effective use of space.

a b

c

Fig. 4 Tourists can rest, play and enjoy the scenery during their visit to the pavilion

2.2 Design Strategy

Diversity of architectural space: sightseeing, landscape, and popular science. This design takes full account of the use function and the particularity of the space, that respects the original properties of the site and the objects of providing services to visitors. A U-shaped progressive fence was designed in the square, where the main building is built by bamboo, quiet and cool. The double-layer U-shaped fence is planted with bamboo of height difference to form a semi-open space. The overall design of the concept is to create a dynamic and static space. The streamlines are curved and straight, combining rigidity and softness. Among them, the U-shaped fence is formed by different bamboo species with introduction tags, that can not only for viewing but also for the scientific popularity of bamboo culture.

Building structure: safe, stable, standardized with high performance. The main raw materials of the building are round bamboo culms, bamboo weaving products, and bamboo crafts. Round bamboo culms are used to support the frame, while bamboo weaving products are used to form an integrated landscape space of viewing and rest. According to the relevant design codes and standards in China, different connection methods are adopted in the structure and the whole structure was analyzed and calculated through the finite element model, so as to meet the requirements of stress, load-bearing, and earthquake resistance of relevant engineering materials. The overall ecology of the building is environmentally friendly, practical, and sustainable.

Organic building environment: integration of architecture and exhibition by using bamboo. The installation space is enclosed by the bamboo structure. The slender and the representative bamboo species are selected to form the bamboo landscape in different heights, which can not only reasonably divide the space to form the narrowed-in scenery and the leaking-through scenery of the traditional Chinese garden, but also meet the growing needs of different kinds of bamboo [3]. Combined with gravel steps, bamboo seats, the bamboo pavilion, and the open-air landscape, we provide a strong line of sight guidance for the environmental space and put on science popularization tags with QR code to facilitate tourists' cognition, so as to truly achieve the interaction between exhibition, tourists and space penetration, see Fig. 5 for details.

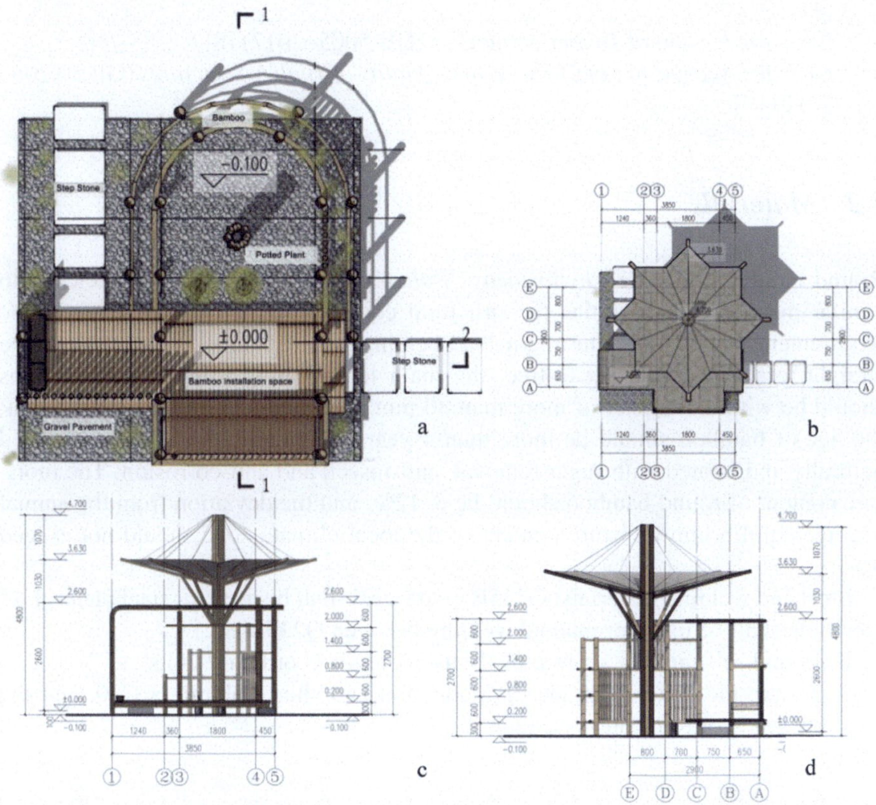

Fig. 5 Plans, sections and elevations. Ground Plan **a**; Roof Plan **b**; 1–1 Section **c**; 2–2 Section **d**

3 Structural Analysis

3.1 Design Criteria

The team adopted the following Chinese Standards and Codes as references:

1. *Load Code for Building Structures* (GB 5009–2012) [4]
2. *Code for Seismic Design of Buildings* (GB 50,011–2010) [5]
3. *Code for Design of Building Foundation* (GB 50,007–2011) [6]
4. *Technical Specification for Round Bamboo Structure Building CECS* 434:2016 [7]
5. *Code for Design of Timber Structures* (GB 5005–2017) [8]
6. *Code for Acceptance of Construction Quality of Timber Structures* (GB 50,206–2012) [9]

3.2 Materials

Round bamboo: Choose *Phyllostachys Pubescens* or other bamboo species with similar mechanical properties for structural components, which should meet the requirements as follows: the main load-bearing round bamboo culms should be straight, crack-free and decay-free; the main load-bearing round bamboo culms should be with a diameter of more than 80 mm and thickness of more than 8 mm; the age of bamboo should be more than 4 years. Bamboo culms should be dried naturally and treated with sugar removal, anti-insect, and anti-corrosion; The moisture content of round bamboo should be 8–12%, and the deviation from the annual average equilibrium moisture content of the local climate zone should not exceed 3%.

Steel and welding materials: Q235B grade steel shall be used for steel plates. E43 electrode shall be used for manual welding between Q235 steel.

Bolts and self-tapping screws: the bolts are grade C ordinary bolts.

Concrete: the concrete grade of foundation and short column is C30, and the concrete grade of foundation cushion is C15.

3.3 Structural Analysis

In Fig. 5, the whole structure can be divided into two independent structural systems. One is the frame structure, which is composed of beams and columns to form a bamboo frame with roof and floor. Another structural system is an umbrella-shaped shelter, which is mainly composed of beams, columns, and cables.

The load of the frame structure includes two parts: one is roof load and the other is floor load. Roof load is transmitted from the roof to purlins, then to beams, columns and finally to the foundation. Floor load is transmitted from floor panel to beams, then to column and foundation. The floor load is 2 kN/m^2, and the roof load is 0.5 kN/m^2.

The roof load of the umbrella-shaped structure is 0.5 kN/m^2. The main frame structure is composed of beams and a single column made of round bamboo. The density of bamboo is 800 kg/m^3. The elastic modulus of bamboo is 12 GPa. The strength of bending and tensile are 60 MPa, for which the safety factor is 3 and the allowable stress is controlled as 20 MPa. The umbrella-shaped structure also consists of steel cables. The design parameters for steel are given as follows: the elastic modulus 206 GPa; shear modulus 79,000 MPa; density 7850 kg/m^3; yield strength 235 MPa; design value of tensile, compressive, and bending strength 215 MPa.

Structural finite element analysis has been carried out for the calculation of the frame structure (Fig. 6) and the umbrella-shaped structure (Fig. 7). The connections of the structure should have reliable strength, and the failure should not before that of the members. Details of different connections are provided in Fig. 8. We also simulate the influence of wind, hot sun, and rain on various indicators of a bamboo building under various environments, to ensure that the building can meet all the requirements in China.

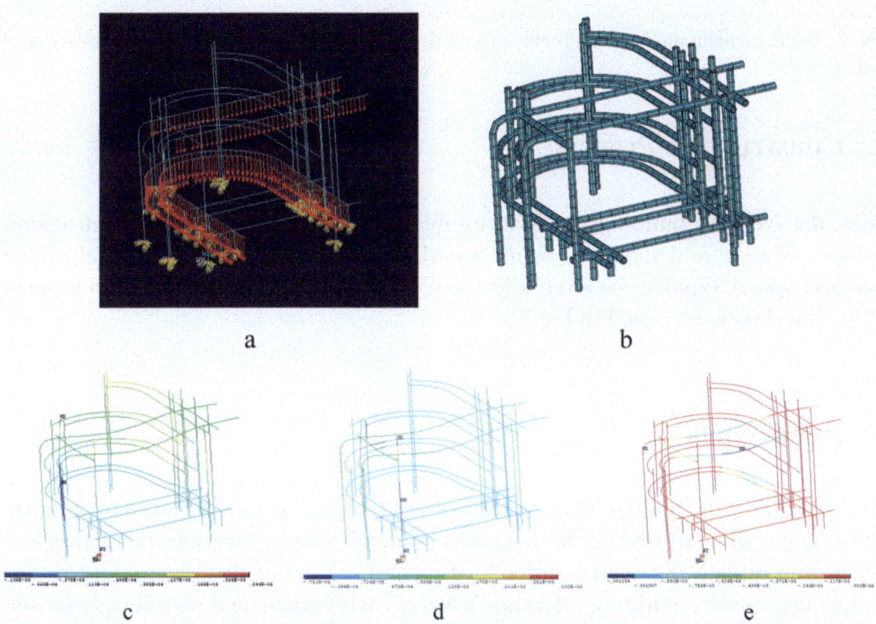

Fig. 6 Finite element model diagram of the frame structure **a**, **b**; X, Y and Z Direction Displacement **c**, **d**, **e**

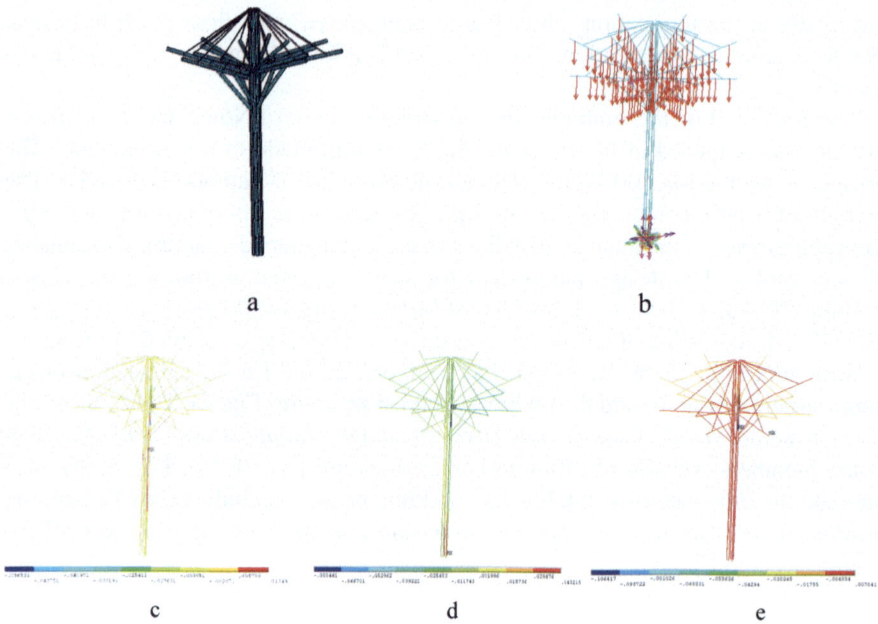

Fig. 7 Finite element model of umbrella-shaped structure **a**, **b**; X, Y and Z direction displacement **c**, **d**, **e**

4 Construction Process

First, the bamboo building needs to establish a foundation, that is, the structural column. The second step is to connect each node. The third step is to enclose the building space. Finally, the overall building is decorated. The construction process of the building is provided in Fig. 9.

5 Renderings

By recognizing the biological characteristics of bamboo, bamboo is used to organize open space, so as to realize the functions of meditation, education, recreation and tea tasting, create the spatial order in the natural bamboo landscape, and finally endow the bamboo building "Bamboo's Verve" with poetic and picturesque artistic conception. The rendering of the design is in Fig. 10.

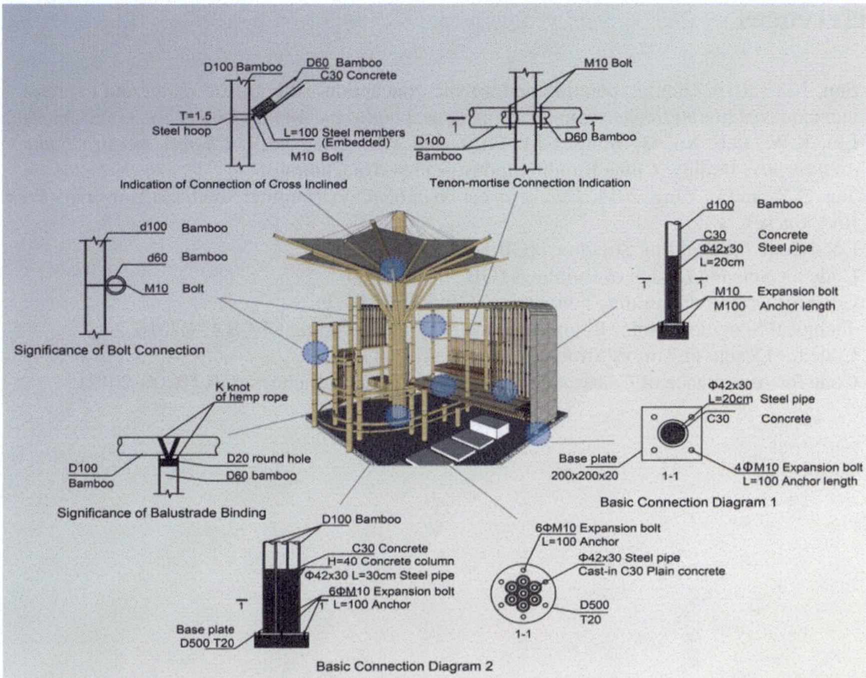

Fig. 8 Details of the different connections

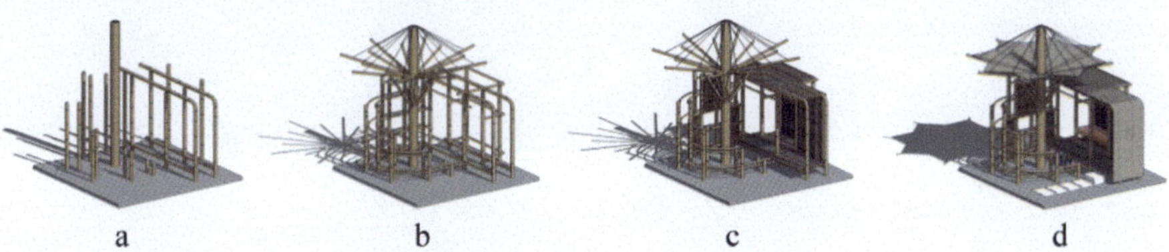

Fig. 9 Construction process: Column **a**; Connect **b;** Enclose **c**; Ornament **d**

Fig. 10 Rendering

References

1. Sun, X.Y. 2013. Habitat, painting and artistic conception—the artistic realm and expression techniques of literati freehand landscape gardens. *Landscape Architecture* 6: 26–33 (in Chinese).
2. Liu, K.W., Q.F. Xu, G. Wang, F.M. Chen, and Y.B. Leng. 2019. *Chinese modern bamboo architecture*. Beijing: China Building İndustry Press (in Chinese).
3. Tan, G.Y., and L. Yang. 2014. *The construction of bamboo*. Nanjing: Southeast University Press (in Chinese).
4. Load Code for Building Structures (GB 5009–2012)
5. Code for Seismic Design of Buildings (GB 50011–2010)
6. Code for Design of Building Foundation (GB 50007–2011)
7. Technical Specification for Round Bamboo Structure Building CECS 434:2016
8. Code for Design of Timber Structures (GB 5005–2017)
9. Code for Acceptance of Construction Quality of Timber Structures (GB 50206–2012)

Island: A Writers' Retreat

David Vardy, Luhang Shi, Haitian Xie, and Xinxin Gao

Abstract The project is imagined as a small temporary bamboo building inside which one may write. The title "Island" refers to the separation of the interior space from the outside of a hot and humid Beijing. This separation is threefold—the physical separation created by a surrounding water-garden whereby the access is made by crossing a bamboo platform; the visual separation of a veil of bamboo textile covering entirely the structure and diffusing light; and a thermal separation achieved by a passive natural ventilation cooling system inspired in part by the barjeel wind-towers of the middle east. A central column of bamboo poles forms a wind-tower. The poles are threaded through a heavy concrete water-filled earth-supported tub at the base. This tub and its steel base plate form a ballast foundation supporting the bamboo super-structure, and act as a water, wind, and earth cooled thermal sink, and as a desk from which to write. The design aims to exhibit the pliability of bamboo; as a construction material; the building being made from bamboo pole, bamboo lath, bamboo-ply, bamboo decking, and bamboo textile. At night, with internal light reflecting in the surrounding water pool, the building is intended to resemble a Chinese paper lantern.

Keywords Bamboo building · Wind-tower · Bamboo structure · Passive ventilation system · Bamboo textile building

D. Vardy (✉)
Wenzhou-Kean University, Wenzhou, China
e-mail: dvardy@wku.edu.cn

L. Shi · H. Xie · X. Gao
Xi'an Jiaotong-Liverpool University, Suzhou, China
e-mail: luhang.Shi15@alumni.xjtlu.edu.cn

H. Xie
e-mail: haitian.xie18@student.xjtlu.edu.cn

X. Gao
e-mail: xinxin.gao15@alumni.xjtlu.edu.cn

K. Liu et al. (eds.), *2019 International Bamboo Construction Competition*, Springer Tracts in Civil Engineering, https://doi.org/10.1007/978-3-030-91990-0_17

1 Concept Design

The project follows an impetus to create a bamboo 'writers' retreat'—somewhere protected and shaded, cooled and still, natural; an inspiring space of concentration and semi-privacy, from which to sit and write.

The inner space is surrounded by a 'reflection-pool', planted with water-lilies and containing floating solar lights. A bamboo platform deck crosses the water to provide access, as well as providing a surface upon which to place a chair or plant pots. Entrance is made by pulling apart a white bamboo textile that envelopes the structure, a curtain of sorts. Inside, there is a cork stool, and a cool-to-the-touch concrete writing desk that holds a volume of water, the surface of which reflects the natural light entering from above.

Bamboo pole framing is formed into a central swirling column and chimney—a wind-catcher (1), orientated to face the general direction of air-flow on a given site. The wind-catcher has a wide opening at the top fitted with a bespoke shaped bamboo funnel, catching and driving the cooler higher wind downward into the space. This fresh air supply is further cooled by the effect of its acceleration, and by passing over a night-time-cooled thermal energy store of concrete and water, at the columns base. The chimney has a central vertical partition that hovers over the waters face, allowing its opposite chamber to operate as a stack extract chimney, drawing out stale warm air, and driving a ventilation circulation through the space.

This naturally ventilated internal environment is conceived as a sanctuary on a hot and humid Beijing day. For an occupant, the effect of the thermal energy store radiating cool; and of the ventilation system passing an accelerated air-flow across the skin; and of the cool-to-the-touch concrete writing desk; each contribute to an improved thermal comfort.

In plan the structure forms a series of swirls and loops, enveloped by a 'veil' of breathable bamboo textile material, acting as a light filter, and resulting in an interior of diffuse daylight. As an adaptation, similar to a tent, fabric sections can be rolled-up and pinned, creating an open pavilion instead of a naturally cooled enclosed one. Configurations of open or closed sections are controlled by the fabric panel design which is aligned to the framing elements, allowing the pavilion to respond to alternative locations and variations in temperature and light throughout the day, and able to open to views or visual connections of other kinds. In this respect the building is adaptive, able to respond to weather and situation, and importantly, able to be adjusted and transformed with speed and ease by a building user (Fig. 1).

Bamboo is among the most versatile construction materials of today, especially with modern innovations and technological processes, and yet perhaps this is not so well appreciated, particularly the potential of bamboo for textiles. The project explores a way to combine a multiplicity of bamboo material products, some traditional construction, and some modern innovations—bamboo-pole, bamboo-lath, bamboo-ply, bamboo-decking, and bamboo-textile—the effect is intended as didactic and unifying. The overall use of bamboo as a material able to achieve sustainable

Fig. 1 Conceptual render

construction is further complimented by an idea of efficiency and a reduction of material use through designed consolidation and multiplicity, meaning elements perform more than one role. For example, the central column is intended to work variously as a wind-catcher, a natural ventilation chimney, a roof light, and a structure.

2 Design Development

The buildings structural frame is made of naturally straight bamboo poles, and also flexed bamboo poles, where slight bending is required of the buildings shape (Fig. 2). This way curvature is achieved almost fully through a simple use of natural bamboo. The one exception to this is the continuous ring beam at the roof eaves. This beam is of a curvature that cannot be achieved by naturally growing bamboo poles, and so a special beam is formed as a bundle of bamboo laths (or splits), glued together and pinned, and forming a prefabricated beam. This beam occurs also as a very visible fold in the fabric that covers the building, and marks the transition from wall to roof, and so a precise curvature rather than a faceted line is important to achieve visually. This prefabricated beam being a curve rather than a faceted line achieves this.

The central column structure is designed as a single-leaf hyperboloid and is constructed using a traditional technique of weaving flexed naturally straight bamboo poles around each other (2). The poles are regularly pinned, and with each pole twice-fixed at the base using proprietary metal brackets. These poles do not touch the ground, and so are kept free of ground moisture. They are also weaved through pre-cast holes formed in the water-weighted concrete tub that acts as a structural ballast, with loads transferred to the ground via a steel spreader-plate attached to

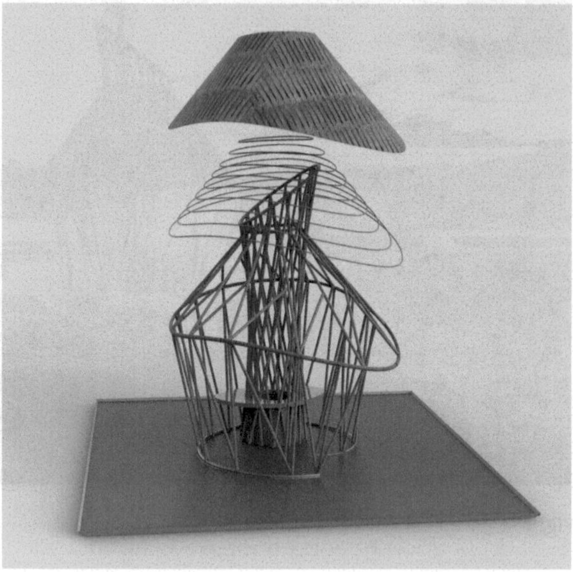
Fig. 2 Structure

the tub. This steel spreader-plate is lipped so as to form and contain the water pool outside, with a precise straight and level edge. The use of a concrete and water weighted ballast and pad as a foundation contributes to the buildings' sustainability, eliminating the need for a traditional invasive and disturbing concrete foundation and under-building. As long as the ground surface is suitably compacted, and the ground is of suitable bearing, the building could be considered a temporary and moveable structure, able to travel between locations with relative ease by way of a process of disassembly and reassembly.

The continuous ring beam at the roof eaves is supported by straight bamboo poles, triangulated and forming a canted wall. The roof is formed as sections of triangulated straight bamboo poles, spanning from ring beam to column and with mid-span support, made rigid by a covering of bamboo lath, pinned and glued to the bamboo poles, and with gaps for light to pass, forming a spiral pattern around the central column, and facilitating water run-off.

The junction between the bamboo poles of the wall and the bamboo lath eaves ring beam, is made by proprietary metal pins and brackets that handle lateral compression and shear resistance forces, and these joints are largely secret fixed with metal work concealed inside of the poles (alternatively there are more traditional options for such joints). The bamboo poles of the wall are fixed into a continuous curved steel angle running the length of the base of the wall and welded to the steel spreader-plate water pool former, which is further weighted by water, river pebbles and rocks, and together with the water-weighted concrete tub structural ballast, this forms a stable and ground-supported structure able to resist lateral forces and uplift (Fig. 3).

The water-weighted concrete tub structural ballast is formed from pre-cast concrete, and as well as being a heat-sink, it also forms a bespoke shaped desk

Fig. 3 Section and elevations

for writing. Cool to the touch, it holds a vat of water that acts along with the concrete itself, as a thermal energy store, cooled at night, and as a drainage system should it rain and water enter the central chimney. As a further use of bamboo, the concrete pre-casting process utilizes a formwork made of bamboo decking boards, such that the concrete is imprinted with the surface pattern and aesthetic of bamboo once struck. These same bamboo decking boards are then reused to form the flooring. This recycling idea and utilization of an element for multiple purposes; such as the tub being a desk, a foundation, and a heat-sink; is a core repeating principal and strategy that contributes to the sustainability of the construction (Fig. 4).

The outer surface of the building is a 'veil', made from state-of-the-art bamboo textile—engineered for impermeability to water droplets, breathable, and with low air-permeability. It allows for the passage of diffuse daylight and is a warm matt off-white color, treated for resistance to mold. This woven fabric can optionally be rolled up to create an open pavilion instead of a naturally cooled enclosed one.

Fig. 4 Floor Plan, Luhang Shi, Xi'an Jiaotong-Liverpool University Design Research Centre (DRC)

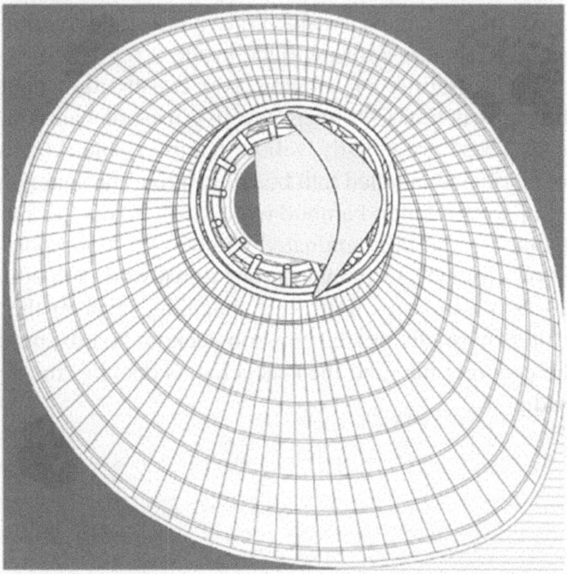

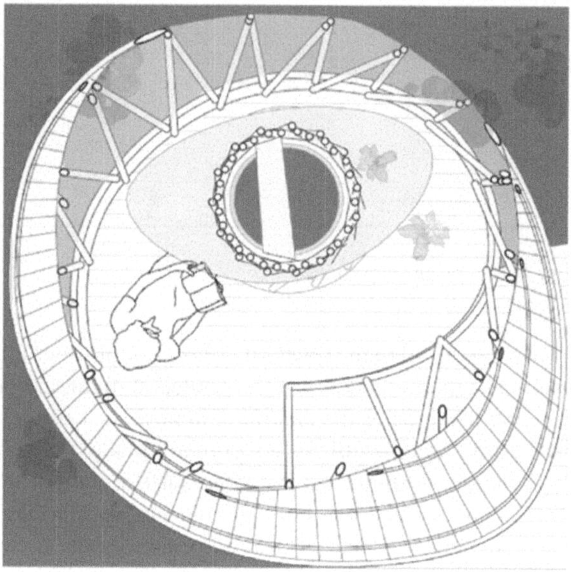

Fig. 5 Floor Plan, Luhang Shi, Xi'an Jiaotong-Liverpool University Design Research Centre (DRC)

Various configurations of open or closed sections can be created following the seams of the fabric panels and their locations, allowing the building to respond to alternative locations, and to variations in temperature and light throughout the day. This adaptability is also a core aspect of the sustainability of the design (Fig. 5).

Writing can be considered a deeply reflective activity. The water garden is formed by a shallow steel liner, with green-grey river pebbles and planted with water lilies and other water plants. Crossing this water is how someone reaches the interior. The platform can be considered as a threshold, marking a journey from one place to another, from the outside to the inside.

The versatility of bamboo as a construction material is employed throughout the construction (3)—there are five uses of bamboo, each exposed and visible to read. These are—firstly yellow bamboo pole, both straight and in the form of a prefabricated bundled lath beam, forming the super-structure; secondly the shower-proof woven fabric bamboo textile that covers the superstructure; thirdly a bespoke double-curved thin laminated bamboo-ply hood positioned at the top of the ventilation chimney to funnel the air downward, something like a roof cowl; fourthly bamboo lath strips used as a sheathing layer at the ceiling above the poles; and fifthly, thermally treated bamboo decking boards, used firstly as formwork to mould the concrete tub, and then recycled as decking for the floor and entrance platform bridge.

3 Experience

Bamboo, as a fast-growing renewable material resource, has the potential to contribute significantly to achieving sustainable construction (4). With this project we endeavored to build not only a bamboo framed structure, but also to use bamboo as flooring, as roofing, and as an envelope material. We designed a sustainable performance in use, through the creation of a natural cooling system that reduces a need for active system energy-consuming cooling, and in so doing we were interested to exploit both the low-tech and the high-tech potentials of bamboo as a material to create a wind-tower, and to create a synthesis and exhibition of these multiple uses of bamboo as part of our small building design.

The architectural idea is concerned with how the control of space, light, temperature, and furniture, can combine to create a space and environment from which an occupant may be inspired in the production of their creative work—in this case writing. Our experience through this project helped us to understand some of the tradition of bamboo form-making, weaving and jointing with bamboo poles, and also how contemporary engineered bamboo can be used to create high-strength bespoke prefabricated double-curved pieces such as the ventilation hood, and surface systems such as textiles. We learned that we could structure and render a building almost fully in bamboo surfaces of one kind or another.

We valued the experience, and learned from the other competition finalists, and from the knowledge and ambitions that now support modern bamboo construction as a field. In our view, the future of bamboo construction can be both a revival or rediscovery of past traditions, and at the same time, the development and innovation of modern high-tech bamboo products and processes, especially as a path to sustainable renewable construction. We hope to have exhibited through our approach and design solution, an example of a small, sustainable, natural, low and high-tech bamboo construction.

4 Rendering

Two renderings are provided as below shown in Fig. 6. The renderings convey an atmosphere of gentle diffused light entering through the surrounding textile and through the skylight of the ventilation chimney, and the reflection of this light by the water surface held by the central writing desk. Other than the concrete (which is formed from a bamboo mould), all surface materials are bamboo, in various forms. The internal space is intended to feel something like being under the canopy of a large old tree; to be a place of serene concentration, somewhere shaded and sheltered, a cooled space from which someone may be inspired to think and to write.

Fig. 6 Render of the interior space

Acknowledgements This project was developed with the support of the Xi'an Jiaotong-Liverpool University Design Research Centre (DRC). The renders are the work of Zeyu Jiao of the DRC Visuals Team, and the drawings are the work of Luhang Shi.

References

1. Fatemeh, J., N. Payam, C. J. Kaiser, M. Yusof, M. Badruddin, S. Salim, S.A. Zaki, H. Ben and Y. Muhammad. 2016. A review on windcatcher for passive cooling and natural ventilation in buildings, Part 1: Indoor air quality and thermal comfort assessment. *Renewable and Sustainable Energy Reviews.*
2. Bambang, K., S. Mohamad, A.Arar, W. Julaihi and S. Bassim. 2020. *Bamboo application in building design: case study of green school*, Bali, Indonesia.
3. Jayanetti, L. and P. Follet. 1998. *INBAR technical report 16: Bamboo in construction—an introduction.* International Bamboo and Rattan Organization: Beijing, China.
4. Manandhar, R., J.H. Kim, J.T. Kim and S. Bassim. 2019. Environmental, social and economic sustainability of bamboo and bamboo-based construction materials in buildings. *Journal of Asian Architecture and Building Engineering*

Re. Source—Retained Source

Giovanni Boccia, Gianluca Lubrano, Valentina Buggè, and Francesco Marmo

Abstract The project takes inspiration from the theme of the contest, the blossoming. In particular, our intent was that of giving the idea of the incipient motion retained in buds. This concept has been rendered in a trivalent way. The most evident interpretation is formal and is represented by the shape given to the structure, which depicts a bud. The mechanical behavior of bent raw bamboo culms that are in the process of releasing their elastic energy and spread upwards is used to symbolize the retained vital energy included in buds. Finally, the building material, chosen to be left almost untouched after harvesting, and the design of structural details, which foster to reuse of structural elements, actualize the reference to the nature given by the contest theme. The design process started from the conceptualization and admiration for raw bamboo culms that are taken as the main and determinant elements of the project. The design choices take advantage of the several aspects connected with the use of this building material such as its strength and weathering, ease of use, assembly, and dismantling costs. We were also concerned about emphasizing the ecological, economical, aesthetic, and mechanical properties of the raw material such as the high specific resistance to traction and bending and the natural cylindrical shape of culms.

Keywords Bamboo culm · Active-bending · Reusable joints

G. Boccia
Department of Civil Engineering, University of Naples Federico II, Naples, Italy

G. Lubrano
Department of Material Engineering and Production, University of Naples Federico II, Naples, Italy

V. Buggè
Department of Architecture, University of Naples Federico II, Naples, Italy

F. Marmo (✉)
Department of Structures in Engineering and Architecture, University of Naples Federico II, Naples, Italy
e-mail: f.marmo@unina.it

K. Liu et al. (eds.), *2019 International Bamboo Construction Competition*, Springer Tracts in Civil Engineering, https://doi.org/10.1007/978-3-030-91990-0_18

1 Concept Design

The main theme of the contest, the blossoming, could be almost automatically linked to the visual idea of the flower, which is colorful, delicate, elegant. But digging in the concept that the word expresses, we can also say that to blossom is an action, which made by the bud (Fig. 1) in order to become what it was always supposed to be. The concept of blossoming recalls a kinetic idea, in perpetual motion, and we decided to play with this interesting concept, in order to express how it beautifully merges with what bamboo itself expresses.

For our project, we focused on the peculiarity of the main material, bamboo. We decided to put under the spotlight the proverbial flexibility of bamboo culms, looking for a design that employs the concept of active-bending [3] and for this reason gives the idea of always being under tension. Also, using the culms would have made our structure modular, easy to assemble with a defined number of similar pieces.

The concept of a tight blossoming relies on the constraining of the bamboo culms with vertical elements, an action that aims to show all the potentiality of bamboo's flexibility in temporary structures (Fig. 2). The natural tendency of the culms to free themselves from the bindings that retain them to expresses the action of blooming in a unique way, like freezing the petals of a flower in the very moment they open up. In this way, the structure is always dynamic in its own essence while being firm and stable.

Fig. 1 The blossoming and idea of the flower

Fig. 2 The concept of a tight blossoming with bamboo poles

2 Design Development

The planform and elevation of the proposed project are shown in Fig. 3. It operates as an open office, warehouse, educational space, or relaxation suite such as a drink corner. A sustainable choice of material, treatments, and joint design provides a long-lasting resistance over time [1].

The structure base is contained in a square of 9 m^2 and is composed of a steel tube curved in such a way to form an open circle. On the superior part of the tube, two groups of tubular pivots, having diameters equal to 4 and 8 cm are used to connect the vertical bamboo elements. Starting from the opening on the left-hand side of Fig. 3, the larger pivots are slightly inclined outwards, while they gradually become vertical in the central part of the base. Smaller pivots are positioned on the top of the remaining portion of the circular base and are all kept vertical.

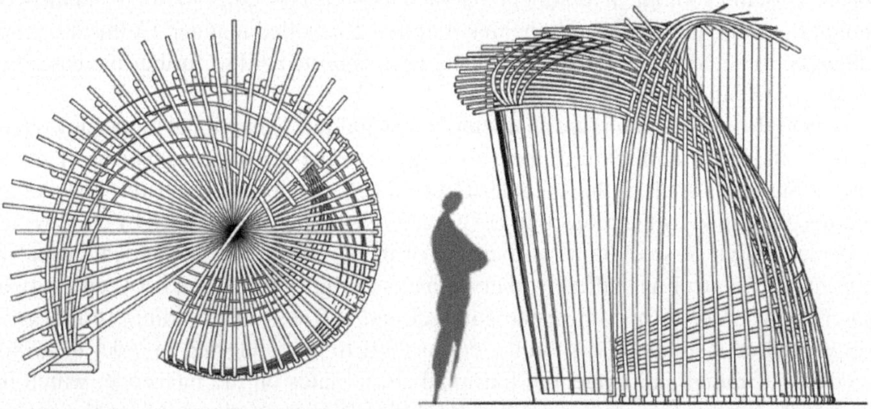

Fig. 3 The concept of a tight blossoming with bamboo poles

Both groups of pivots serve as the basis for bamboo culms. Larger culms (D80—t8), which are stiffer, are kept straight, until they reach the top of the structure, at about 4 m from the base. Thinner and more flexible culms (D40—t8) are active-bent and cross over the planform to reach the top of larger poles [3]. A series of diagonal bamboo stripes are joining the larger and straight culms, in order to improve the stiffness of the structure. Curved bamboo stripes are used to transversally join the curved thinner culms. The top of the structure is covered with additional bamboo stripes and branches in order to furnish shading and protection yet allowing natural ventilation. Bamboo branches have excellent insulating properties, both from a thermal and acoustic point of view [5, 6].

The choice of the most suitable material for the structure has been dictated by several considerations including the limits imposed by the contest, material strength and weathering, ease of use, assembly, and dismantling costs. We are also concerned about emphasizing the ecological, economical, aesthetic, and mechanical properties of the raw material such as the high specific resistance to traction and bending and the natural cylindrical shape of culms. Other considerations were made for the sustainability of our project. The employment of raw material favors a reduced amount of transformation processes that the material needs to undergo before being used in the construction industry. Obviously, this limits the ecological footprint of the project in terms of CO_2 equivalent emissions related to the building process.

It must be said that the best choice to meet these resolutions would have been that of selecting a particular species of bamboo, that is *Dendrocalamus Strictus*, characterized by larger wall thickness and better capability to support larger flexural deformations. However, this choice was not possible due to the contest restrictions. For this reason, keeping the idea of using natural bamboo culms, we decided to prescribe methods and treatments that increase their duration so as to foster its reusability. Among the various treatments that can be used for bamboo, the most sustainable ones regard a proper selection of the harvesting period of stems and the so-called curing, a totally ecological treatment that increases the resistance to fungi and insects and, at the same time, dries out the material in a controlled manner. Culms are then subjected to an additional treatment that employs boric acid to further increase the resistance to fungi.

Although bamboo culms are characterized by unique, exceptional properties, such as strength, form, and texture, they also inherit some drawbacks, typical of most natural materials such as the irregularity of their geometry, which barely meets the tolerances defined by the project. In our case, this issue has been solved by employing a special design of steel joints at the base of the structure. These elements usually require bold control by skilled workers to assess a tight connection between the two materials in order to maximize the strength and stiffness of the joint. To this end we prescribed a suitable finish at the culms ends to be connected in order to avoid the onset of radial and tangential tensional shear states on the internode, which is weaker, that can put the connection to a hard test. The connection between the culms to the base is realized through an interlocking pivot joint welded onto the base tube (Fig. 4). The connection between each culm and the base is made by interlocking male–female elements which are then fixed by a bolt mechanism (Figs. 4 and 5).

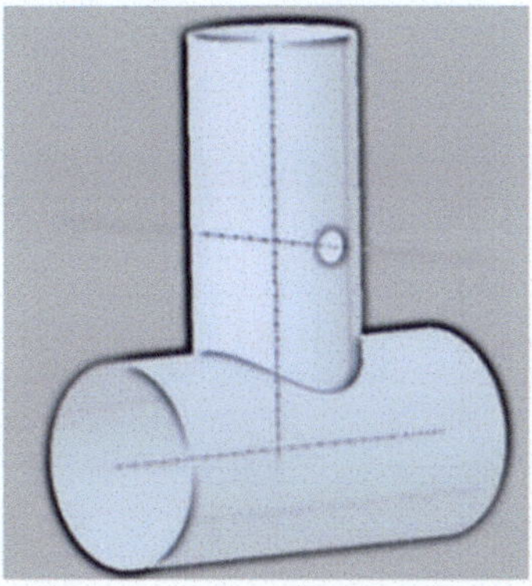

Fig. 4 Pivot joint between the base tube and the vertical stems

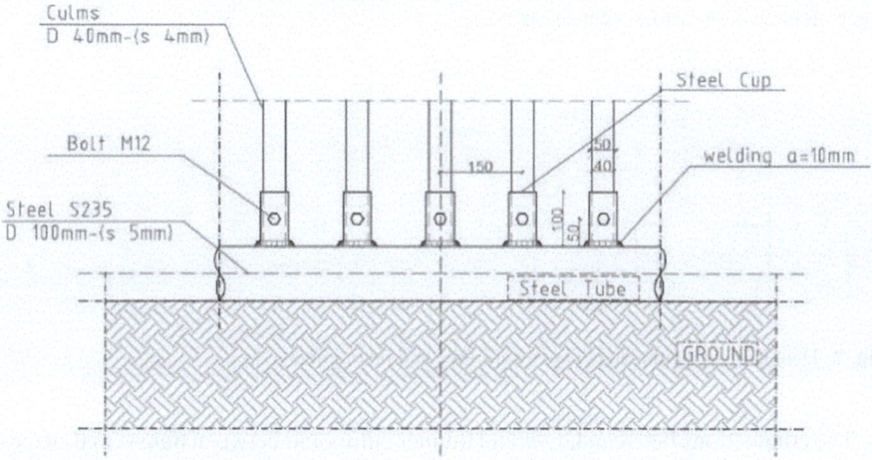

Fig. 5 Base pivot-joints for 4 cm culms

Here, nodes closer to the connections are trimmed in order to regularize the form of joining parts, which avoids undesired local stress concentrations and increases the adhesion with the support.

To further increase the adhesion to the support, a butyl rubber sealant cover is prescribed for the part of the stem that will enter the joint. This technique also avoids issues related to possible stagnation of water inside the joint and the contact with the end of the culm which may weaken the connection over time [1, 2, 4]. Accordingly, the thickness of the pivot joint is prescribed to be at least 2 mm. Additional constraints connected to the weldability of the pivot forced us to prescribe a thickness of 5 mm (Figs. 6 and 7).

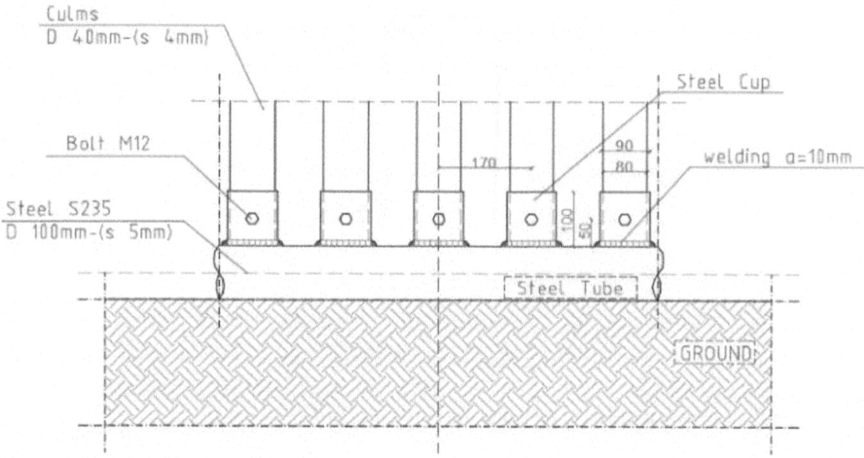

Fig. 6 Base pivot-joints for 8 cm culms

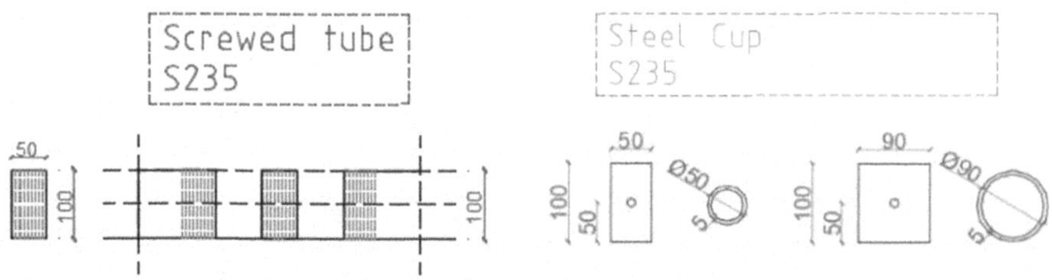

Fig. 7 Details of screw thread between base tube and pivot joints

The connections between larger and thinner culms and between transversal stripes and both orders of culms are made by knots of natural bamboo rope or by the interweaving of bamboo stripes, for which a thermal treatment is prescribed to increase their flexibility. Knots are positioned in such a way that each damaged el-ement can be easily substituted without compromising the stability of the structure. To ensure higher adhesion between knots and bamboo culms we prescribed a spe-cial treatment for the area in the vicinity of each joint. The connection area has to be cleaned and covered with bi-adhesive butyl rubber sealing tape. This ensures the tightness of the connection due to increased surface friction between surfaces. It al-so improves normal actions between surfaces in contact due to the elasticity of the rubber, which fosters a tighter manual assembly.

In order to verify the deformability and strength of the proposed project, a finite element model of the structure has been employed. The model has been implemented and analyzed within the finite element software Nolian AllInOne (Softing srl, Rome, Italy). Bamboo culms have been modeled by employing beam elements fixed at the base and connected with each other through pinned joints.

The finite element analysis allowed us to verify both the deformability and strength of the structure. According to the numerical simulations, the most stringent requirement is related to deformability checks. To this end, we assumed a limit to the lateral displacements at the top of columns that is proportional to the overall height of the building so as to meet usability and comfort requirements. This limit has been set to one-hundredth of the height of the building.

Figure 8 shows a contour plot of displacements produced by a force of 1000 N, applied at the top of the lateral openings of the lateral walls, which resulted in the most vulnerable points of the entire structure versus the deformability requirements.

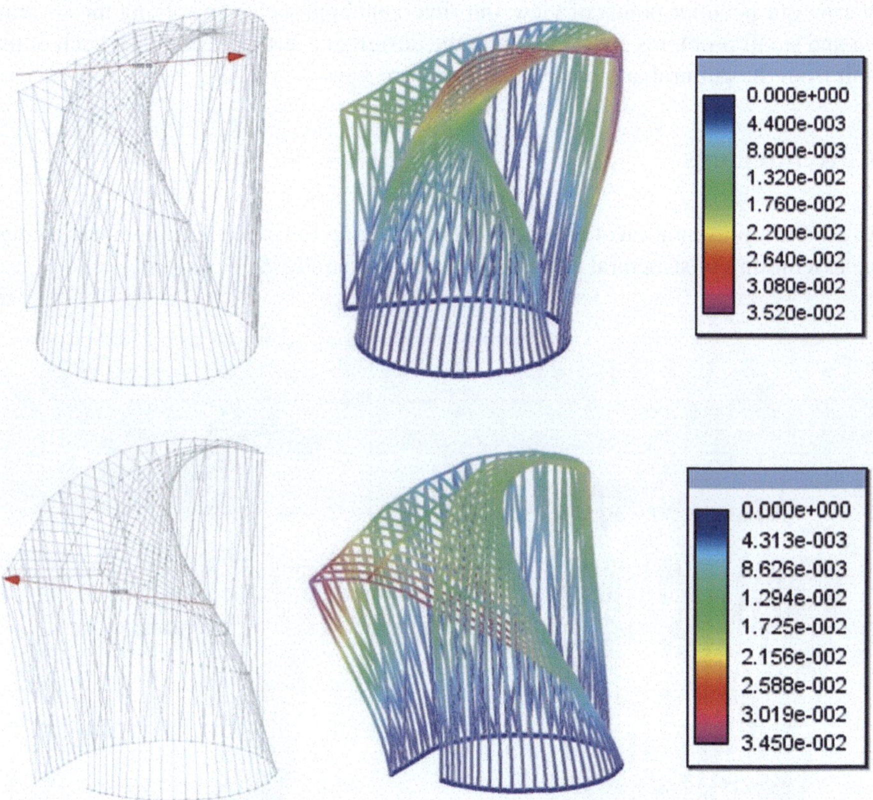

Fig. 8 Displacements produced by a horizontal force applied at the most vulnerable points of the structure

3 Experience

The design of a small bamboo structure has served as a priceless formative experience both for the students and the tutor. The entire team had the occasion to deepen their knowledge and consciousness about the potential of bamboo as a building material. None of us had previous experience with this unique material, which we learned to appreciate for its mechanical, aesthetic, and ecological properties.

The teamwork has been very instructive and stimulating. The three students had different experiences since we came from diverse and disjoint formative paths, related to Building Engineering, Material Engineering, and Architecture. The possibility of sharing our peculiar points of view, our divergent approaches in solving the several big and small problems connected with the design of a structure enriched each of us both from the cultural and professional point of view.

4 Renderings

An artistic representation of the proposed structure and an exploded diagram showing the assembling of structural elements are reported in Fig. 9.

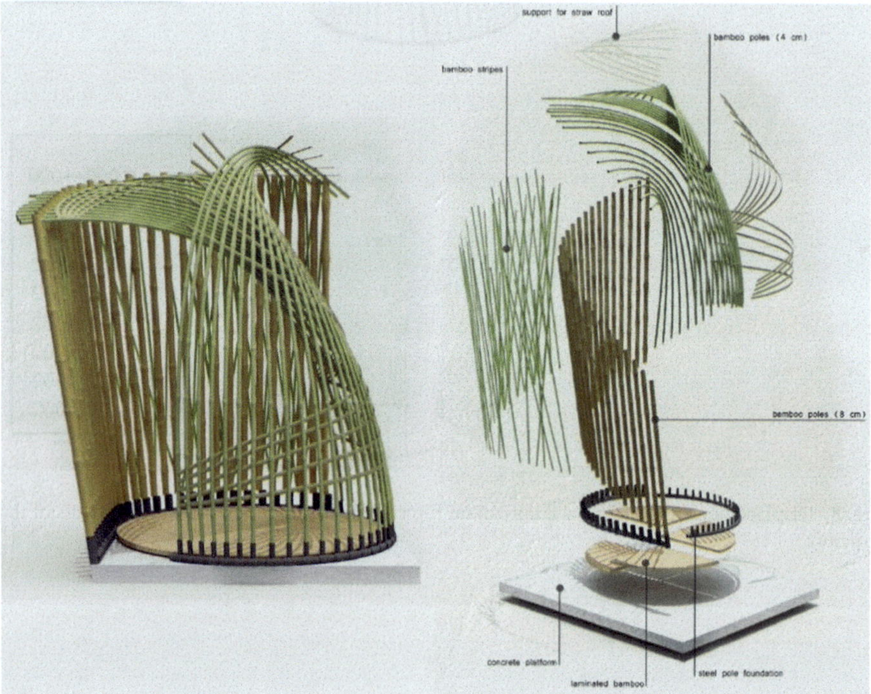

Fig. 9 Artistic representation of the structure and exploded view of the structural elements

References

1. Arce-Villalobos, O. A. 1999. Fundamentals of the design of bamboo structures (Doctoral dissertation, Technische Universiteit Eindhoven).
2. Jayanetti, D. L., and P.R. Follett. 2008. Bamboo in construction. In Modern bamboo structures, pp. 35–44. CRC Press.
3. Lienhard, J., H. Alpermann, C. Gengnagel, and J. Knippers. 2013. Active bending, a review on structures where bending is used as a self-formation process. *International Journal of Space Structures* 28 (3–4): 187–196.
4. Xiao, Y., M. Inoue, and S. Paudel (eds.). 2007. *Modern bamboo structures.* CRC Press, Taylor & Francis Group.
5. Zhou, S.C., C. Demartino, and Y. Xiao. 2020. High-strain rate compressive behavior of Douglas fir and glubam. *Construction and Building Materials*, 258: 119466.
6. Wang, J.S., C. Demartino, Y. Xiao, and Y.Y. Li. 2018. Thermal insulation performance of bamboo-and wood-based shear walls in light-frame buildings. *Energy and Buildings* 168: 167–169.

Golden Flower

Paola Arcos, Katherine Tintín, José Calvo, and Enrique Mauricio Barreno Avila

Abstract Traditional constructions are generally realized using metal, adobe, cardboard or concrete structures. In this study, we consider to build a structure by using *phyllostachys edulis*, a Chinese bamboo species also known as "Moso bamboo". Bamboo, as a new trend in the construction market, is a good acoustic and thermal insulator. Compared to traditional construction materials, bamboo is light, easy to handle, and with very good earthquake-resistance performances; in fact, it has been used in certain regions of Ecuador and Colombia by using different anchoring and fastening techniques. In this project, we implement mooring and anchoring techniques, which are traditional construction approaches. The proposed pavilion is an external bamboo installation made of six petals. Based on principles of dynamics, colour, size and proportion, an innovative bamboo structural design was created after the required calculations. The design is inspired by the concept of golden ratio. Golden Flower, an architecture installation by using round bamboo poles and strips, fully showing the material and artistic characteristics of bamboo.

Keywords Eco-friendly material · Aesthetic design · Bamboo installation · Golden ratio

1 Introduction

Bamboo is a plant of great importance for rural areas inhabitants [1]. Bamboo has an immense potential to be an environmentally friendly material, not only due to its

P. Arcos (✉) · K. Tintín · J. Calvo
Faculty of Architecture and Urban Planning, Universidad Central del Ecuador, Quito, Ecuador
e-mail: pcarcos@uce.edu.ec

J. Calvo
e-mail: jecalvo@uce.edu.ec

E. M. Barreno Avila
Faculty of Civil and Mechanical Engineering and Technology and Transfer Center (CTT), Universidad Técnica de Ambato, Ambato, Ecuador
e-mail: ebarreno3907@uta.edu.ec

K. Liu et al. (eds.), *2019 International Bamboo Construction Competition*, Springer Tracts in Civil Engineering, https://doi.org/10.1007/978-3-030-91990-0_19

astonishing mechanical properties, flexibility and great resistance for the elaboration of furniture, musical instruments or tools, but also due to its large diameter which is suitable for processing into structural components of houses and boats [2].

Being a light material, bamboo is easy to transport, store, facilitate, build quick for the construction of temporary or permanent structures in different terrains [2]. Furthermore, this plant can be used after some transformations to make various construction materials, such as mats, strips, handmade or industrialized laminated bamboo products. Bamboo, therefore, has become an important alternative for various construction projects, due to its multiple uses and its rapid growth [2].

The material required by the Organising Committee of IBCC 2019 is moso bamboo (*Phyllostachys edulis*), which is often harvested in China during the winter [3]. The height of the plant varies between 12 and 15 m and its diameter is between 6 and 13 cm. We will use this bamboo species to design our pavilion. The proposed pavilion is an external bamboo installation made of six petals. The design is inspired by the concept of golden ratio.

2 Design Process

2.1 Concept Design

The concept of the Golden Flower was born from the idea of the golden ratio. In mathematics, two quantities are in the golden ratio if their ratio is the same as the ratio of their sum to the larger of the two quantities. In geometry, a golden spiral is a logarithmic spiral whose growth factor is the golden ratio. That is, a golden spiral gets wider (or further from its origin) by a factor equal to the golden ratio for every quarter turn it makes. An approximation of a golden spiral is a Fibonacci spiral, which is constructed slightly differently. A Fibonacci spiral starts with a rectangle partitioned into 2 squares. In each step, a square the length of the rectangle's longest side is added to the rectangle. Let us take the Fibonacci spiral as a module of our design. A set of six spirals are rotated on the plane present seen in Fig. 1. Thus, the design shape of the Golden Flower comes to life in a configuration provided by the central circle of the 6 spirals, repeating in synchronization of its axis and becoming an inscribed hexagon. Around the central hexagon, six petals are also abstracted as the final shape of the design.

2.2 Design Development

During the design development, the whole structure is divided into two parts: one is the upper hexagon structure surrounded by 6 petals, which was inspired from the concept of the golden ratio; the other is a central supporting structure, consisting of

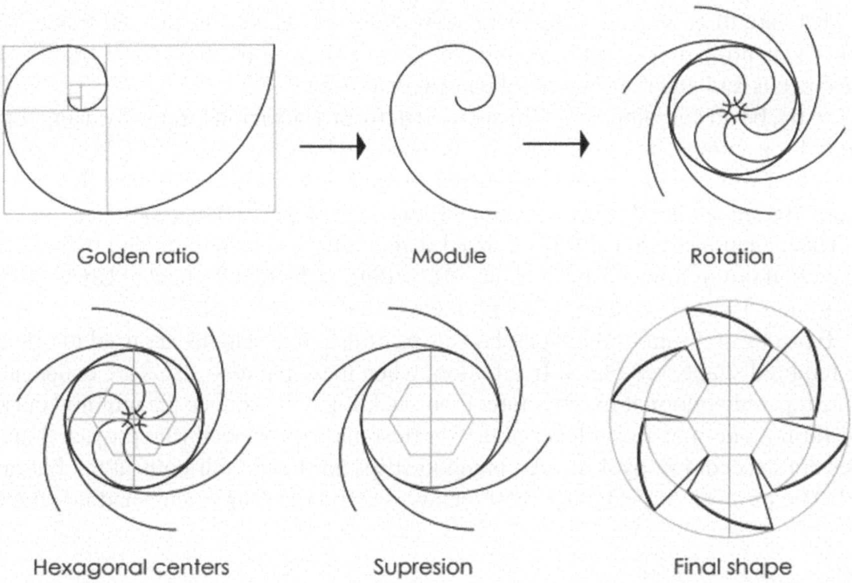

Fig. 1 Golden Flower's shape process

6 round bamboo culms. Figure 2 illustrates the top, elevation and section view of Golden Flower.

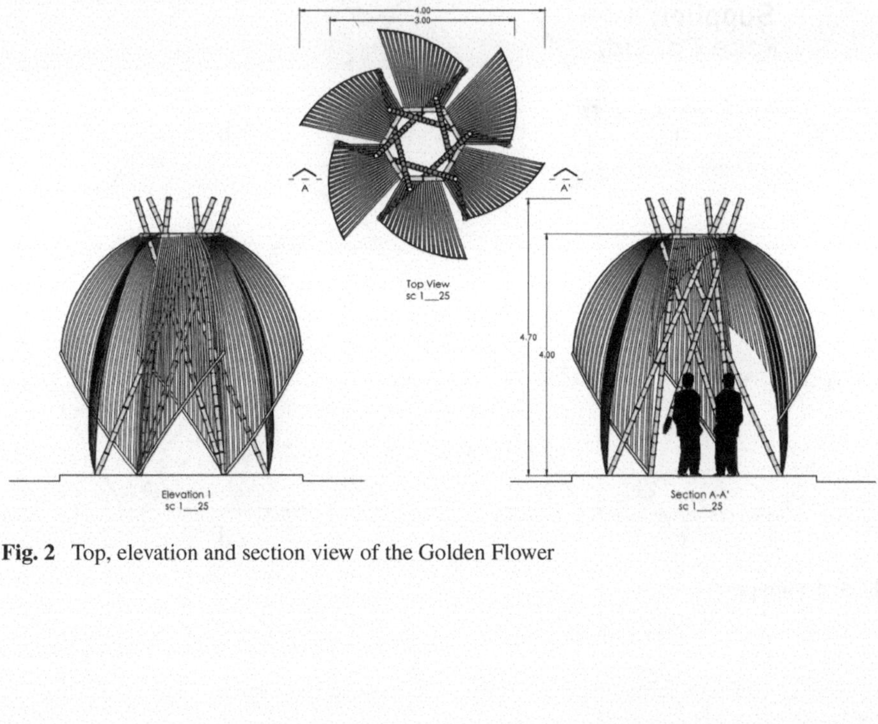

Fig. 2 Top, elevation and section view of the Golden Flower

There are three steps to construct the structure: (1) make a hexagonal frame; (2) form the central supporting structure and connect it with the hexagonal frame; (3) make petals and attach them onto the hexagonal frame.

Firstly, bevel cut joints (Fig. 3b) are used to form a hexagonal frame by using rope and bolt fastenings. The hexagonal frame needs to bear the weight of falling petals.

Secondly, frame the central supporting structure. Six bamboo culms with a diameter of 80 mm are fixed to the hexagonal frame at its upper end by ropes (see Fig. 3c, d). Then, each bamboo culm is arranged diagonally, and its bottom end is fixed on the ground through steel bars and bolts. According to the requirement of IBCC 2019, the ground floor needs to be kept within the area of 3 m by 3 m.

20 pieces of 3 mm of thick bamboo strips of different lengths are used to obtain the required shape, see Fig. 4. In addition, when these bamboo strips are connected to the top and bottom of the structure, their buckling will produce the required curve and form a one-meter cantilever at the shortest part of the petal (the projected area does not exceed 4 m by 4 m). All bamboo strips are fixed with bolts at the bottom end. The purpose of designing these petals is to produce light and shadow effects

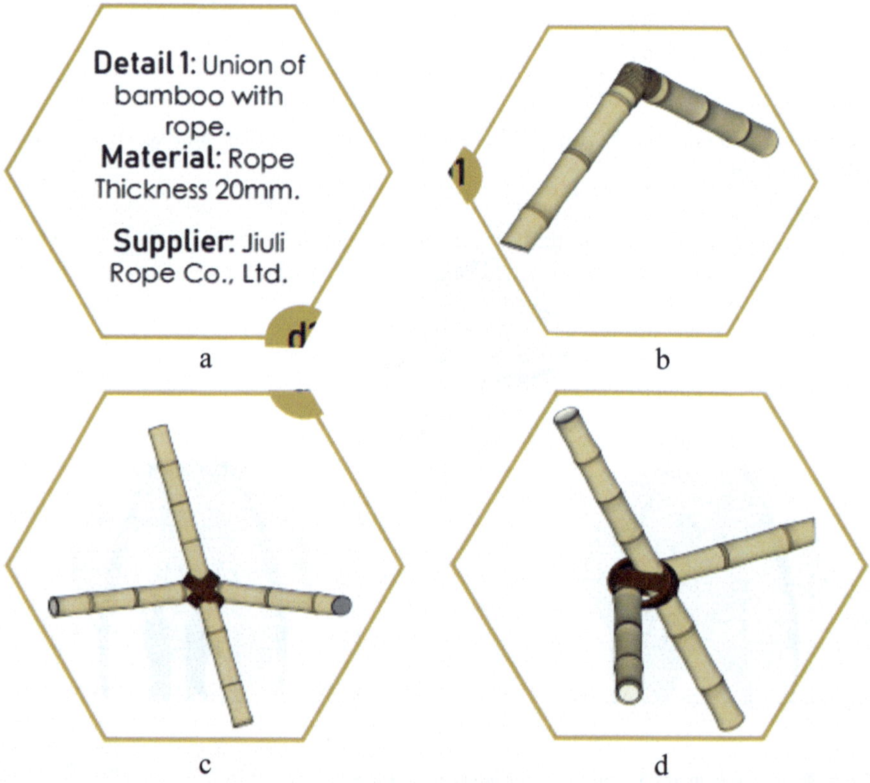

Fig. 3 Joint types

Fig. 4 Making petals

in the process of their interleaving, as shown in Fig. 2. After the main structure is formed, all the petals can be placed on it (Fig. 4).

The spaces are created by petals making users to access freely. At the same time, the installation also has permanent ventilation and cooling effect, which helps to realize the main goal of this design.

3 Discussion

The proposed project (Golden Flower) aims to expose and publicize the current uses of bamboo as a building material in contemporary and modern architecture and to replace traditional high-energy consumption construction materials, like concrete and steel.

In this design, we followed the appropriate use of bamboo. We use round bamboo poles as columns, which are supported by a concrete base on the ground. Thus, the end of the bamboo will never be in contact with the ground directly, avoiding bamboo corrosion, even when it is outdoors.

4 Renderings

Several renderings of Golden Flower are provided in Fig. 5. The renderings are showing the pavilion with some visitors in the structure and providing a schematic

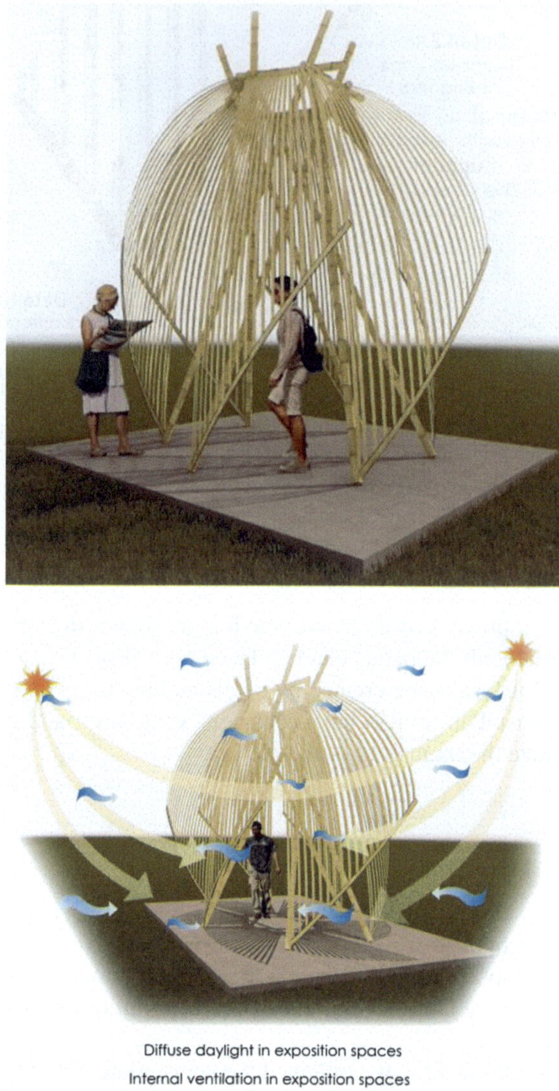

Fig. 5 Renderings of Golden Flower

of the ventilation of the exposition space. Humans are added to the renderings to facilitate the visualization of the scale of the photos.

References

1. Phimmachanh, S., Z. Ying, and M. Beckline. 2015. Bamboo resources utilization: A potential source of income to support rural livelihoods. *Appl Ecol Environ Sci* 3 (6): 176–183. https://doi.org/10.12691/AEES-3-6-3.
2. Vintimilla Salas, E.G. 2020. Bamboo connected: Rethinking bamboo connections: designing a simplified system for connecting bamboo in order to develop housing units for a determinate target population in Ecuador. Last accessed in 29 July, 2021. https://repository.tudelft.nl/islandora/object/uuid%3A4b9f8822-8f60-4f17-85d2-c8726fb6eacb.
3. Zhao, H., et al. 2018. Chromosome-level reference genome and alternative splicing atlas of moso bamboo (Phyllostachys edulis). *Gigascience* 7 (10): 1–12. https://doi.org/10.1093/GIGASCIENCE/GIY115.

Blossom Business

Camilo Andrés Moreno, Denis Alejandra Sepúlveda, and Jaime Andrés Valenzuela

Abstract This project aims to meet the need for housing through a structure that optimizes space and creates an environment of easy access, construction, and affordable based on three pillars: (1) The first is the design, which is inspired by the conceptual structure of a flower with clean cuts and a minimalist concept. It has two spaces that are complementary: the first floor is an office and the second a meeting room that can function as a study room; (2) The second is the structural verification, thus the structural analysis is based on the Colombian Earthquake Resistant Standard which has a specific design section with bamboo *Guadua Angustifolia Kunth*; (3) The third is the use of natural and renewable raw materials in all the structure and finishes, such as wood and bamboo, thus achieving one hundred percent potential of these two materials.

Keywords Bamboo · Versatile · Blooming · Accessible · Efficient · Sustainable constructions · Ecological materials in construction

1 Concept Design

1.1 Basic Scheme

The design is based on a simple abstraction of a flower. First, we take its basic structural elements (stem and branches) to incorporate them into the structural design and formal composition of space. Second, we take the most beautiful moment of its vital process when it is flowering to dispose of multiple windows on the ceiling and

C. A. Moreno · D. A. Sepúlveda · J. A. Valenzuela (✉)
Department of Civil and Agricultural Engineering, Universidad Nacional de Colombia, Bogota, Colombia
e-mail: jvalenzuelat@unal.edu.co

C. A. Moreno
e-mail: cmorenoe@unal.edu.co

D. A. Sepúlveda
e-mail: dasepulvedad@unal.edu.co

K. Liu et al. (eds.), *2019 International Bamboo Construction Competition*, Springer Tracts in Civil Engineering, https://doi.org/10.1007/978-3-030-91990-0_20

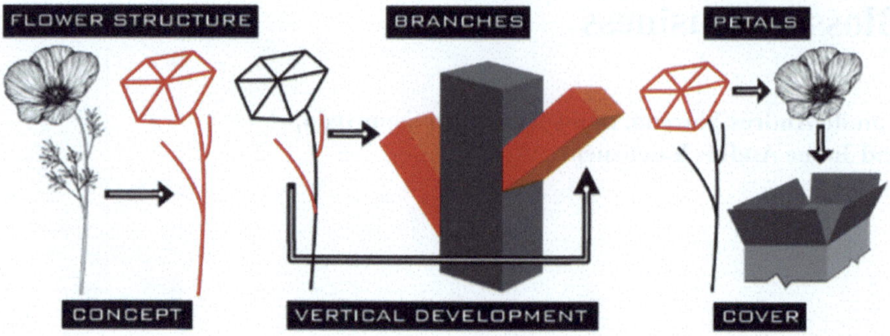

Fig. 1 Design concept.

facade as the abstraction of the petals, as shown in Fig. 1. Applying clean cuts with a minimalist concept, it was possible to come together two different spaces: an office on the first floor and a complimentary space as a meeting room or optimized study space on the second floor.

1.2 Environment Adaptation

At present, the need for space due to the fast growth of cities is becoming important in architecture to propose solutions for its use and maximization, using simple, functional, and economic designs. For this reason, we propose a construction with traditional techniques and unconventional or natural materials to facilitate its accessibility to the common population.

The environment in which you will find the structure and the summer climate in which it will be, it is around 23–28 °; a reason why the structure presents a separation of mezzanines of 2.7 m that allows the natural recirculation of the air and the second floor is arranged as a mezzanine that leaves a large space to avoid the accumulation of hot air inside the enclosure. This also raised the strategy of using a double coverage for the structure, made of tissue mad; so that there is an "air cushion" between the walls that will act as isolation to avoid direct contact from inside with the outside as is shown in Fig. 2.

The total cost of the proposal considering the structure, finished in detail, and tools necessary for the construction is 4991 USD. We try to make it economical so that it can be a viable and attractive alternative this is shown in Tables 1, 2, 3 and 4.

Fig. 2 Ventilation System.

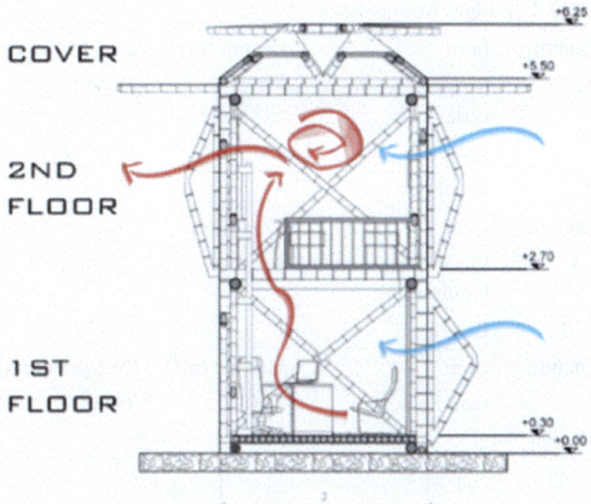

Table 1 Budget for structure

Quantity	Item	Length (m)	Cost per unit (COP)	Reserve	Total (COP)
50	Bamboo	6	27,000	5	1,485,000
124	1/2-inch rod	0.5	5100	7	668,100
248	1/2-inch nuts	N/A	281.7	14	73,797
248	1/2-inch washers	N/A	281.7	14	73,797
In total					2,300,693

Table 2 Budget for tools

Quantity	Item	Length (m)	Cost per unit (COP)	Reserve	Total (COP)
2	Drill	N/A	1,000,000	–	2,000,000
4	spanner	N/A	106,900	–	427,600
2	chainsaw	N/A	900,000	–	1,800,000
2	Disc for metal	N/A	10,000	–	20,000
2	Disc for wood	N/A	50,000	–	100,000
4	Broach	N/A	20,000	–	80,000
In total					4,427,600

Table 3 Budget for finishes

Quantity	Item	Length (m)	Cost per unit (COP)	Reserve	Total (COP)
27	Strips of white cedar	2.5	21,000	2	609,000
1	staircase	4.1	2,400,000	0	2,400,000
23	Bamboo	6	27,000	3	702,000
Quantity	Item	Area (m^2)	Cost per meter (COP)	Reserve	Total (COP)
20	Floor in laminated bamboo	1	227,800	1	4,556,000
120	Estera	1	11,429	6	1,440,000
Quantity	Item	Volume (m^3)	Cost per unit (COP)	Reserve	Total (COP)
1	wood glue	0.0038	35,000	–	35,000
In total					9,742,000

Table 4 Budget

Total (COP)	Total (US)
16,470,293	4991

2 Design Development

2.1 *Volumetric Development*

The architectural volume has a displacement of 1 m on the second floor, which generates a vacuum inside the space, used as circulation and connection between the architectural floors; to the outside of the building creates a cantilever where decorative vegetation will be hung through a ledge. The first floor has a dimension of 3 m × 3 m, 9 m^2 in total, and the second floor 4 m × 4 m, a total of 16 m^2 as shown in Fig. 3.

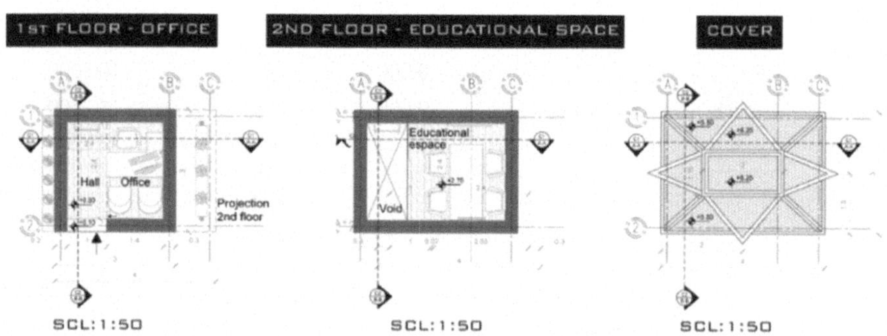

Fig. 3 Architecture floors and cover.

2.2 Spatial Development

In both spaces, the management of the windows is relevant to guarantee natural lighting and ventilation, which can provide a comfortable work or study environment, without requiring many energy-related expenses and they are distributed as follows: the first space is the office, consisting of the main hall and an office of 5 m^2 (2 m × 3 m) with one workplace and two public attention chairs; the second space is a bigger educational space, designed to allow group work or with a large amount of material. It has a group work table, white chairs, and a TV screen for show materials (Fig. 3). These spaces are connected with a reclining metallic ladder, embedded in the structure.

2.3 Skin Cover

The *Guadua Angustifolia Kunth* structure is lined with a woven mat to the exterior and interior of the building in the form of a double skin, whose function is to thermally isolate the interior environment, reducing the temperature and making the occupation of the place more pleasant. This design has multiple customization possibilities since the implemented mat can vary in shades of green-yellow and brown colors. There are also various fabrics to create elements in one piece (Fig. 4).

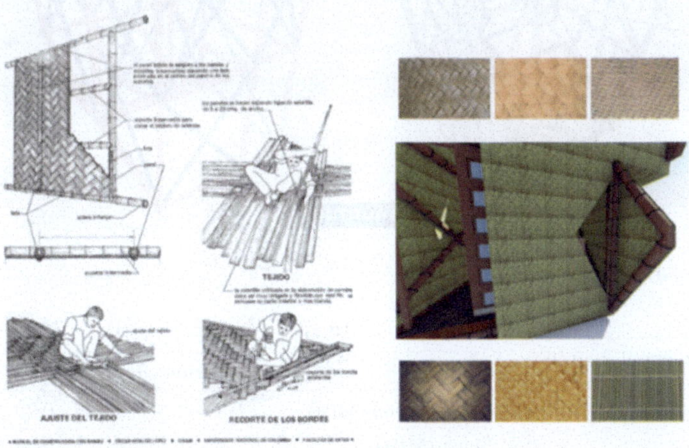

a [1] b

Fig. 4 Mat issue management.

2.4 Structural Design

The structural design was carried out in accordance with the requirements of the Colombian Earthquake Resistant Standard (NSR-10), we have taken advantage of the fact that this norm contains a title with general requirements of design to use wood and *Guadua Angustifolia Kunth* as construction materials [1], that allows us to guarantee the viability of the construction of the structure and additionally to place it in a real context.

It should be noted that thanks to the high flexibility of the material, the structural analysis will be governed by the control of the permissible deflections, always checking the bending, crushing, and cutting resistance.

In this way, after a load analysis, a model was made using SAP2000 V21.1.0 (see Fig. 5), this model allowed us to perform the corresponding verifications. It is important to note that following the title G of NSR-10, the admissible deflections must respect the ratio L/360 for mezzanines and L/240 for roofs.

Thus, Table 5 shows that the values of admissible stresses and deflections are adequate to guarantee the proper functioning of the structure [2].

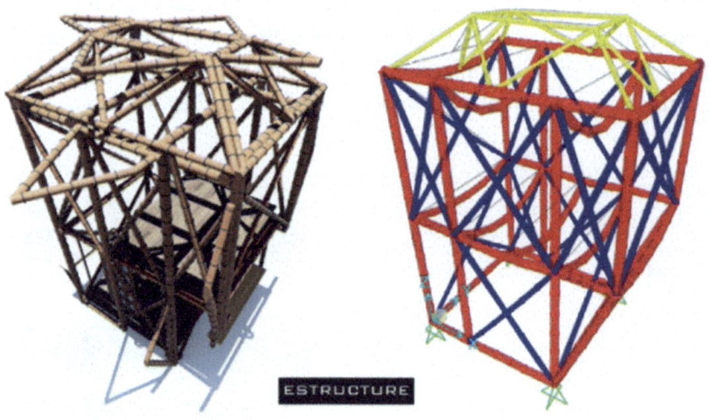

Fig. 5 Structural design

Table 5 Structural analysis

Strength	Permitted	Max
Compression (MPa)	17	0.29
Shear (MPa)	4	0.98
Tension (MPa)	8	0.012
Deflection (m)	0.0083	0.00091

3 Experience

We entered the contest with high expectations, the team approached this project with great professionalism. A team composed of two civil engineering students and an architecture student make up the flourishing business project.

This idea began by defining the needs that it wanted to meet. In this case, variables such as the optimization of space, the environmental impact or carbon footprint generated by the structure were carefully considered, the combination of environments that it should have to complement the space in the best way, the value of the building, etc.

Once all this was defined, the team proceeded to do an investigation of the subject to have quality information and thus generate the base design on which to work. So when the first challenges began to appear around this stage, the joint work was crucial for carrying out the project, the fact of carrying out the architectural and structural design at the same time served to get a structure of clean cuts, optimized in spaces, stable and at the same time that ideas arose such as using "mat" as a "skin" of the structure which would serve two purposes: one, to serve as a thermal insulator given the high temperatures that can arrive in different zones in summer, and the second, provide a design similar to the flower petals in the windows.

Then it goes to the structural evaluation stage where it is modeled in the SAP2000 program to verify that it meets the standards stipulated in the NSR-10, but the new challenge arose is the way to model the structure so that it would be as close as possible to the reality, by getting used to designing with isotropic, linear and uniform materials such as concrete or steel, taking into account variables that can affect the model, for this we used research that yielded important data such as the maximum length that can be worked on an element considering a uniform diameter, the types of joints that can be used, the mechanical properties of the material, among others.

Once the structural design was finished, the team continued with the cost part where the main challenge was to know the cost of raw material and labor in the place to develop the product, possibly in this section a more in-depth investigation had to be done to achieve a more realistic price, however, with the information that was available at the time, an estimate was taken and published.

In conclusion, the contest greatly stimulated the team to generate good interdisciplinary group work, overcome those barriers generated by the lack of information that we have, and promote the use of these valuable and useful materials that nature gives to us.

4 Renderings

Several renderings of the structure are provided, as shown in Figs. 6, 7 and 8.

Fig. 6 Inside view of the structure on the second floor.

Fig. 7 Inside view of the structure on the first floor.

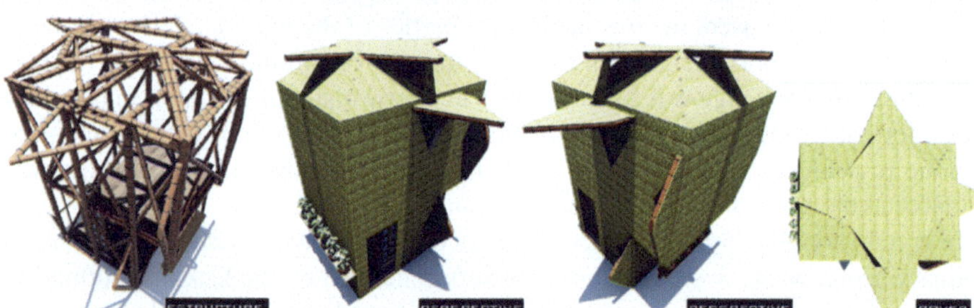

Fig. 8 Structure in different perspectives.

References

1. Asociación Colombiana de Ingeniería Sísmica. 2010. Normas Colombianas de Diseño y Construcción Sismo-Resistente, NSR-10. Bogotá, AIS.
2. Yu, H. Q., Z.H. Jiang, C.Y. Hse, and T.F. Shupe. 2008. Selected physical and mechanical properties of moso bamboo (*Phyllostachys pubescens*). *Journal of Tropical Forest Science* 258–263.
3. Obataya, E., P. Kitin, and H. Yamauchi. 2007. Bending characteristics of bamboo (Phyllostachys pubescens) with respect to its fiber–foam composite structure. *Wood Science and Technology* 41 (5): 385–400.

Conclusions and Future Perspectives of IBCC 2019

Kewei Liu, Cristoforo Demartino, Zhi Li, Qinghui Liu, and Yan Xiao

Abstract The International Bamboo Construction Competition 2019 (IBCC 2019) is the first-ever international competition focusing on bamboo construction, encouraging students of Architecture, Civil Engineering, and Landscape related majors to explore the potential of bamboo as a construction material. For IBCC 2019, approximately 500 students from 30 countries participated in the competition and submitted 70 designs in total. Four of 15 finalists realized their design at INBAR Garden and others presented their design concept to the jury. Based on those articles shared by students in the book, this chapter summarizes four key outputs and three global impacts of IBCC 2019 from the perspective of the organizing committee and opens a discussion on future perspectives.

Keywords IBCC 2019 · Student competition · Bamboo construction competition · Training

K. Liu · Q. Liu
International Bamboo and Rattan Organisation, Beijing, China
e-mail: kwliu@inbar.int

C. Demartino (✉) · Y. Xiao
Zhejiang University - University of Illinois at Urbana Champaign Institute, Zhejiang University, Zhejiang, PR China
e-mail: cristoforodemartino@intl.zju.edu.cn

Department of Civil and Environmental Engineering, University of Illinois at Urbana-Champaign, Urbana, IL 61801, USA

Z. Li
College of Civil Engineering and Architecture, Zhejiang University, Hangzhou 310058, Zhejiang, PR China

K. Liu
School of Civil Engineering, Tsinghua University, Beijing, China

K. Liu et al. (eds.), *2019 International Bamboo Construction Competition*, Springer Tracts in Civil Engineering, https://doi.org/10.1007/978-3-030-91990-0_21

1 Introduction

As a gift of nature, bamboo is closely related to people's lives with extensive uses in "clothing, food, shelter, and transportation". In the context of global climate change, bamboo can greatly help carbon storage no matter as a live plant or becoming a durable product. If bamboo products are used to construct houses, carbon could be sequestrated for 50 years or longer. Coupled with the good performance of its mechanical properties, bamboo is undoubtedly a good substitute for other higher-carbon materials, such as concrete and steel in the construction sector (described in Li et al. Use of Bamboo in Constructions; Simi et al. A Review on Round Bamboo Structural Applications and Perspectives).

However, a limited number of companies and research groups across the world understand how to use bamboo in the right way to build houses. Train young students of Architecture, Civil Engineering and Landscape-related majors from across the world and encourage them to explore the great potential of bamboo as a construction material is the major purpose for INBAR and its partners to organize IBCC 2019 (Liu et al. The 2019 International Bamboo Construction Competition). By the chance of 2019 Beijing Horticultural Expo, global students with different cultural backgrounds gathered in INBAR Pavilion (detailed description in Laverde INBAR Garden Pavilion: Bamboo Eye), the largest round bamboo arch structure in northern China, and competed for their ideas of how to use bamboo to build a temporary garden house after several months of "learning by doing". To compete winning prizes, four of 15 finalists realized their design at INBAR Garden and others presented their design concept to the jury.

After the competition, the organizing committee of IBCC 2019 invited all 15 finalists, as well as several teams with distinctive designs but not award-winners to write scientific articles to share their ideas in detail, including concept design, design development, construction phase as well as their valuable experience during the whole competition. Although it is the first time for most of the students to use bamboo in their design and some of their ideas are still immature, the experience during the whole participation process is very valuable and worth recording. The Organizing Committee sincerely hopes that the words and pictures reported in this book can bring enlightenment to more students, tutors as well as institutions who will organize similar competitions in the near future.

In this chapter, we will summarize key outputs and global impacts of IBCC 2019 from the perspective of the Organizing Committee and open a discussion on future perspectives of this context.

2 Key Outputs of IBCC 2019

Around 500 students from 30 countries registered for the IBCC 2019. The Organizing Committee finally received 70 designs from 12 countries in total. The teams are from

China and Ecuador were the most, accounting for 44 and 29% of the total respectively. Teams from China, Ecuador, Ethiopia, Colombia, Costa Rica, Brazil, Mexico, and India showcased that bamboo is not strange to them as those countries are rich in this plant. There are also several teams from countries with few bamboo resources, such as the United States, Australia, Belgium, and Italy.

Among all the teams who submitted their design, 93% are from INBAR's member countries. For countries where INBAR Headquarters (China) or regional offices (Ecuador, Ethiopia, India, Ghana, and Cameroon) are located, there are a large number of participating teams. There are two reasons for the geographic distribution of students' participation as follows: on the one hand, there are effective ways for INBAR to publicize the competition in these countries through its long-term partnership; on the other hand, INBAR has been carrying out bamboo related projects in those countries for a long time, that has accumulated a lot of data and experience for the utilization of local bamboo resources for students' reference. A large amount of INBAR Technical Reports in English, Chinese, French, and Spanish are available on INBAR's website and can be accessed for free. Most of the teams in this book cited INBAR's technical report as references.

Four major outputs of IBCC 2019 are presented in the following.

2.1 Improved Students' Understanding of Bamboo as a Construction Material and the Whole Construction System

Most of the students have never touched bamboo in their class before participating in IBCC 2019. In fact, the majority of higher education and vocational training institutions across the world do not provide training for building professionals, such as carpenters, engineers, architects, or designers on bamboo [1], due to bamboo is not one of the mainstream construction materials in the commercial market. However, after a few months' "learning by doing", the students who participated in IBCC 2019 successfully completed the design and converted their design concept into real houses by using bamboo. They studied in detail the characteristics of bamboo in its architectural performances, such as its physical and mechanical properties (tension, compression, bending, and shear), durability, and type of connections (by metal joints or lashing). At the same time, they also established a framework understanding of construction systems in a short time, including material preparation, construction process (foundation preparation, set up the structural frame, compete for enclosure and decoration), and construction technologies. While exploring the potential of bamboo as a construction material, the knowledge the students got from IBCC 2019 can also be easily extended to the study of other materials and other structural systems.

2.2 *Stimulated Students' Creative Inspiration*

Due to the wide geographic distribution of bamboo, its architectural applications are strongly influenced by regional cultures, styles, and ideas. Bamboo can greatly stimulate students' creative inspiration. For IBCC 2019, we received quite different designs under the same theme. For example, the team of "House Loti", composed of three members from Belgium, China and the UK respectively, drew inspiration from traditional bamboo and rattan crafts and created a very elegant Asian-style lotus-shaped house in an ingenious way (see Mertens et al. House Loti). "Machine of Bamboo" was designed by a Chinese team, who prefers the concept of modern industrial construction. Dozens of small-sized engineered bamboo components were connected by hundreds of bolts, making their pavilion be produced like a machine (see Erkissa et al. In-Box House). However, the team of "Ternary Shed" from Australia created a house integrating more diverse cultures. They combined the traditional round bamboo design techniques from Latin-American culture, the modern concepts of timber design applied in Australia, and modern engineered bamboo materials created in China (see Wu et al. Machine of Bamboo).

2.3 *Provided a Fair Opportunity for Global Students*

As described above, 93% of participating students are from INBAR's member countries. Currently, INBAR has 48 state members, most of which are developing countries, where bamboo is one of the effective tools for them to improve their livelihood development. IBCC 2019 provides a fair opportunity for students from developing countries to compete with those from developed countries. The competition is completely open to global students without any limitations. Any student from Architecture, Civil Engineering and landscape related majors can participate in the competition, no matter which country they are from or what language they are speaking. At the same time, the jury committee consists of 23 jurors from 11 countries, covering Africa, the Americas, Asia, and Europe. Therefore, the equity and transparency of the competition are guaranteed.

From the final result, we are particularly pleased to see that two of three participating teams from Ethiopia were shortlisted among the top 15 finalists. The team of "in-Box House" was very outstanding, winning the top three. They well combined its national housing policies in Ethiopia, chosen the most innovative engineered bamboo material for creating a very modern home for local people, which can also fit for commercial activities to increase incomes. During their participation in IBCC 2019, they met several barriers to travel to China to complete their construction. INBAR helped them obtain visas through the help of the Chinese Embassy in Ethiopia. Finally, three young students got the chance to complete the construction of "in Box House". As the team described in Erkissa et al. In-Box House, "*Not only IBCC 2019 brought us a great opportunity to share ideas, brainstorm and learn from the rest of the finalist teams, but also gave us a chance to experience a new culture and lifestyle. Our stay in Beijing and at the International Horticultural Exhibition 2019 was an eye-opener that helped us know more about different cultures, lifestyle, building systems and architecture.*" Their experience of participating in IBCC 2019 will reflect in their coming study, work and life, which will definitely help them to serve Ethiopia better in the future. It is the significance of why INBAR organized the IBCC 2019.

2.4 Trained Students to Showcase Their Ideas in a Scientific Way

Writing a scientific paper to share ideas and experiences during the competition is another challenge proposed by the Organizing Committee to the invited teams. It is not easy, especially for undergraduate students who have not been trained in school for this specific task. It is also a great opportunity to practice writing skills. After all, the paper is written to record their own works.

Firstly, students need to record the whole process logically in detail. We provide all the teams the basic frame of the article, including concept design, design development, construction phase, and experience. Secondly, articles need to be expressed in a scientific way with a fixed format provided by Springer Nature. Some students misunderstood the abstract and background while some of them don't know all the figures, tables, or references that should be cited in the text. Thirdly, students from non-native English-speaking countries also have many problems with scientific and technical writing.

Therefore, this book took editors more than one and a half years for collecting and reviewing all the articles by several turns: less than one-third of articles meet the requirement when they were submitted at the first time, among which a few are with high-quality. One-third of the articles were returned back to authors to revise with major revision. The remaining third was largely revised by editors. Nevertheless, all the editors of this book are still fully confident in the quality of this book, because we are often moved by their sincere emotion and serious attitude between the lines.

3 Global Impacts of IBCC 2019

3.1 Impact on Construction Sector

Mr. Cui Kai, Fellow of Chinese Academy of Engineering, Honorary President and Chief Architect of China Architecture Design & Research Group was invited as the Chairman of the jury committee, and 22 senior international construction experts were invited as the jury members. The activity was supported by 40 relevant institutions around the world, including relevant international organizations, industry associations and universities, research institutions, and production and processing enterprises, which laid a good foundation for the dissemination of the activity in the industry. As the activity was organized in China, a number of local media have released a series of reports on the competition, including China Green Times, World Architecture, Landscape Architecture, World Bamboo and Rattan, and etc. Feedbacks from many global experts indicated that "bamboo" is particularly suitable for students to be selected as the subject of curriculum design. Its easy-to-use characteristics can greatly exercise the practical ability of students from relevant majors. IBCC 2019 provides a useful path for the further exploration of capacity-building of global bamboo construction professionals.

3.2 International Impact

IBCC 2019 was first unanimously recognized by the INBAR Construction Task Force. Around 30 senior bamboo construction experts from 18 countries helped the organizing committee send invitations to colleges and universities in their countries from the beginning, which helped colleges and universities in various countries form teams to participate in the competition. In addition, the Organizing Committee also made use of INBAR's platform to release the information, including the official website of INBAR in Chinese and English, Microblog, Twitter, Facebook, LinkedIn, and other network platforms. A series of publicity and reports have been carried on the event lasting half a year, which has attracted a large number of participating teams all over the world. The creativity and vitality of students' design works have inspired a large number of global followers, that greatly contributed to the development of the global bamboo construction industry.

3.3 Public Impact

The communication path for the public mainly includes on-site display and media publicity. Many tourists came to visit INBAR's Bamboo Eye every day during its exhibition. From June 2019, 15 finalists' design works began to be exhibited in INBAR Pavilion, and public voting for the best popular work award was opened. In July, four realized bamboo houses were completed for tourists. The whole exhibition, lasting more than 100 days, attracted the attention of at least hundreds of thousands of tourists. In addition to the continuous sharing and push of the competition by industry institutions, the competition used the Flash Live Broadcasting platform to broadcast the Final Expert Review Meeting and the Award Ceremony, and the number of viewers on the Flash Live Broadcasting platform exceeded 26,000. Nearly 20 local mainstream media in China reported on the award ceremony of the competition, including China Global Television Network (CGTN), China Daily, and People's CN, which pushed the activity to a climax. These reports have a wide audience coverage and attracted the attention of tens of thousands of ordinary people, which makes the competition not only popularize the cognition and skills of bamboo architecture professionals but also achieve the purpose of popular science. IBCC 2019 makes the artistic beauty of bamboo architecture and the concept of low-carbon environmental protection deeply rooted in the hearts of the people.

4 Future Perspective

4.1 Bamboo's Potential as a Training Material in Construction Related Courses

IBCC 2019 not only improved students' understanding of bamboo as a construction material but also helped students understand what the whole construction system is. Compare to other traditional materials like concrete and steel, bamboo is much easier to use. No need for large mechanical equipment, even electricity, anyone can create various kinds of functional spaces by using bamboo with simple tools. Through very simple processes like cutting, splitting, weaving or bending, bamboo could be used to build different components of the house in various forms: straight columns, curved beams, ceilings, windows and etc. For IBCC 2019, four students can complete a small-scale bamboo house in 24 h (net time). It is an excellent material for a short-term design course. Currently, China, Italy, Indonesia and many other countries begin to set up short-term courses on bamboo architecture. Combined the point of view in Sects. 2.1, 2.2 and 3.1 of this chapter, we recommend that bamboo can be widely used for students of Architecture, Civil Engineering and Landscape related majors to carry out small-scale construction practical exercises in school, especially in those countries rich in bamboo species for construction.

4.2 Stimulate Scientific Exploration of Bio-Material for Construction

Learning bamboo is also challenging due to its anisotropy, which could stimulate students' scientific exploration of bio-material for constructions. Especially in the context of climate change, the exploration of this direction has great significance. Natural bamboo is a unidirectional fiber-reinforced composite material (Li et al. Use of Bamboo in Constructions; Simi et al. A Review on Round Bamboo Structural Applications and Perspectives). Cellulose acts as reinforcement, similar to steel bars in reinforced concrete, with its distribution decreasing from outside to inside in the cross-section. This nature distribution of cellulose offers a better stiffness to the bamboo culm which is an excellent example of the structural design acumen of Mother Nature [2]. A series of scientific problems need to be explored by more young people. For example, how to grade bamboo and produce products with stable mechanical properties, how to use appropriate constitutive relations to describe different kinds of natural bamboo or various emerging engineered bamboo products, or how to improve the fire resistance and durability of bamboo products. Only with a full understanding of the before-mentioned problems, bamboo can be more safely used in buildings and play a greater role in our daily life.

4.3 Strengthen Multilateral Collaboration to Support Further Development of IBCC

As described in 3.2 and 3.3 of this Chapter, relied on INBAR's long-term partnership in 48 member countries and the support from the expert team of INBAR Construction Task Force, IBCC 2019 received the full support of dozens of institutions around the world at the beginning of the event. Meanwhile, local national media in China and social platforms of INBAR played a very important role in advertising the context. Based on the basic budget provided by INBAR and Zhejiang University, the Organizing Committee also got fully supports from several local enterprises in China, providing bamboo materials and part of financial support. Therefore, organizing such a competition requires multilateral collaboration, including extensive partnerships, effective publicity channels, and sufficient financial support. To encourage more students to participate in future competitions on bamboo constructions, the best way is to set up long-term relationships with universities all over the world, and make the competition become one of their short-term courses in which students can get credits. In this way, professors can really integrate teaching into competition guidance and encourage students to show themselves on such an international platform.

5 Conclusions

IBCC 2019 is the first-ever international competition focusing on bamboo construction, encouraging students of Architecture, Civil Engineering, and Landscape related majors to explore the potential of bamboo as a construction material. Approximate 500 students from 30 countries participated in the competition and submitted 70 designs in total. Four of 15 finalists realized their design at INBAR Garden and others presented their design concept to the jury. Based on those articles shared by students in the book, this chapter summarizes four key outputs of IBCC 2019: (1) Improved students' understanding of bamboo as a construction material and the whole construction system; (2) Stimulated students' creative inspiration; (3) Provided a fair opportunity for global students; (4) Trained students to showcase their ideas in a scientific way. IBCC 2019 also has had some positive impacts, including the impact on the construction sector, the international impact, and the public impact. In the future, we recommend that bamboo can be widely used for small-scale construction practical exercises in school which can stimulate scientific exploration of bio-material for construction, especially in those countries rich in bamboo. An extensive partnership, effective publicity channels, and enough financial support are important for the further development of IBCC.

6 Photos of IBCC 2019

A series of photos of the Jury Review Meeting and the Awarding Ceremony of IBCC 2019 are provided as follows from Figs. 1, 2, 3, 4, 5, 6, 7, 8, 9, 10, 11, 12, 13, 14, 15, 16, 17, 18, 19 and 20.

a. Jury Review was held on 19 July, 2019 at INBAR Pavilion.

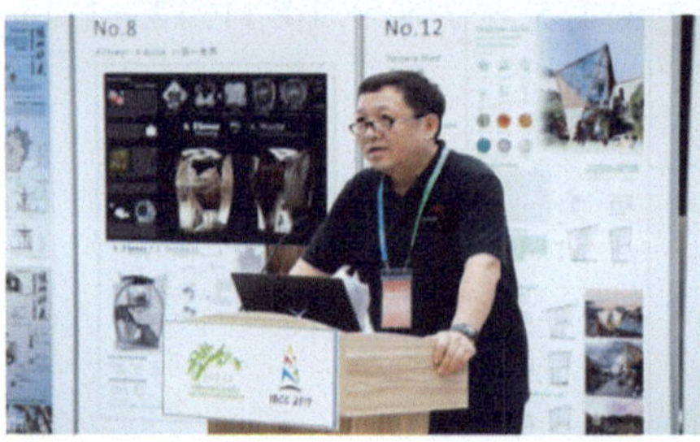

b. Review rules interpretation by Mr. Cui Kai, Chair of the Jury of IBCC 2019, Fellow of Chinese Academy of Engineering, Honorary President and Chief Architect of China Architecture Design & Research Group.

c. Introduction to IBCC 2019 by Ms. Liu Kewei, Chair of the Organizing Committee of IBCC2019, INBAR Global Bamboo Construction Programme Coordinator.

d. Li Zhi, Co-Chair of the Organizing Committee of IBCC2019, Associate Professor of Zhejiang University.

e. Jury Review Meeting moderated by Dr. Cristoforo Demartino, Assistant professor, Zhejiang University and UIUC.

Fig. 1 Kick off of the Jury Review Meeting of IBCC 2019

a. Team of "House Loti" from Belgium.

b. Team of "Machine of Bamboo" from China.

c. Team of "In-Box House" from Ethiopia.

Fig. 2 Presentations delivered by 15 finalists during the Jury Review Meeting

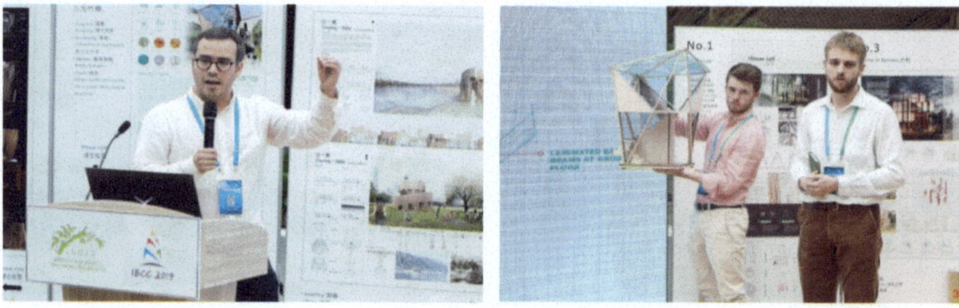

a. Team of "Ternary Shed" from Australia.

b. Team of "A Flower · A World" from China.

c. Team of "Crouching · Hidden" from China.

Fig. 3 Presentations delivered by 15 finalists during the Jury Review Meeting

a. Team of "Bamboo Field" from China.

b. Team of "Bloom" from China.

c. Team of "Island" from China.

d. Team of "Push and Pull Varying Tranquility Bamboo Tectonic" from China.

Fig. 4 Presentations delivered by 15 finalists during the Jury Review Meeting

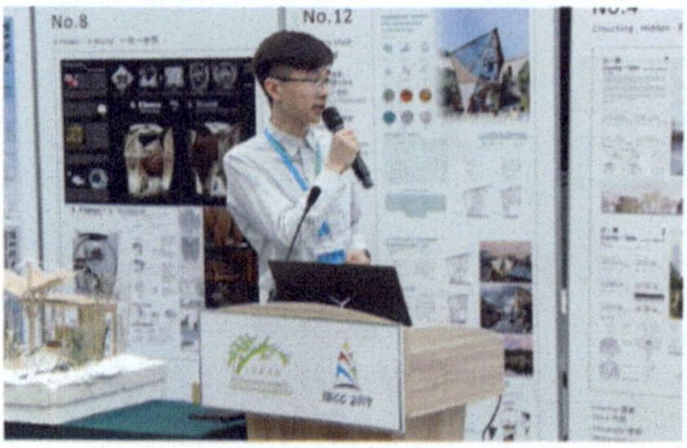

a. Team of "Bamboo's Verve" from China.

b. Team of "Nomad" from China.

c. Team of "Confronting the Performance of Bamboo in Architecture" from Ethiopia.

d. Team of "Bud Pavilion" from China.

e. Team of "Bending Blossom" from China.

Fig. 5 Presentations delivered by 15 finalists during the Jury Review Meeting

Fig. 6 Onsite jury members raise questions to students during the Jury Review Meeting

Fig. 7 Top 3 finalists introduce their demonstrated projects on site

a. Ali Mchumo, Director Genearl of INBAR is delivering opening speech.

b. Mr. Cui Kai, Chair of the Jury of IBCC 2019, Fellow of Chinese Academy of Engineering, Honorable President and Chief Architect of China Architecture Design & Research Group is delivering speech.

c. Yan Xiao, Changjiang and Expert Distinguished Professor at Zhengjiang University-University of Illinois Joint Institute is delivering speech.

d. Asaye Alemayehu, Chargé d'affaires of the Ethiopian embassy in Beijing is announcing the Winner team of IBCC 2019 .

e. Prof. Zhang Li, Dean of School of Architecture, Tsinghua University is announcing the Winner of Public Vote.

f. Prof. Zhao Chen from Nanjing University is announcing the Winner of the 3rd Prize.

Fig. 8 Awards Ceremony of IBCC 2019

a. First Prize: House Loti, designed by a team from the Université de Liège. (Team: Audrey Mertens, Fantine Fontaine, Chen Qi; Advisor: Vincent Denoël)

b. Second Prize: Machine of Bamboo, designed by a team from Nanjing University. (Team: Huimin Wu, Xiaonan Li, Sicheng Zhou; Advisor: Xiao Fu)

c. Second Prize: In-Box House, designed by a team from Addis Ababa University. (Team: Lelissa Erkissa, Mikyas Tekle, Simon Mucheye; Advisor: Boja Muluneh)

Fig. 9 Awarding First Prize and Second Prize to Top 3 teams

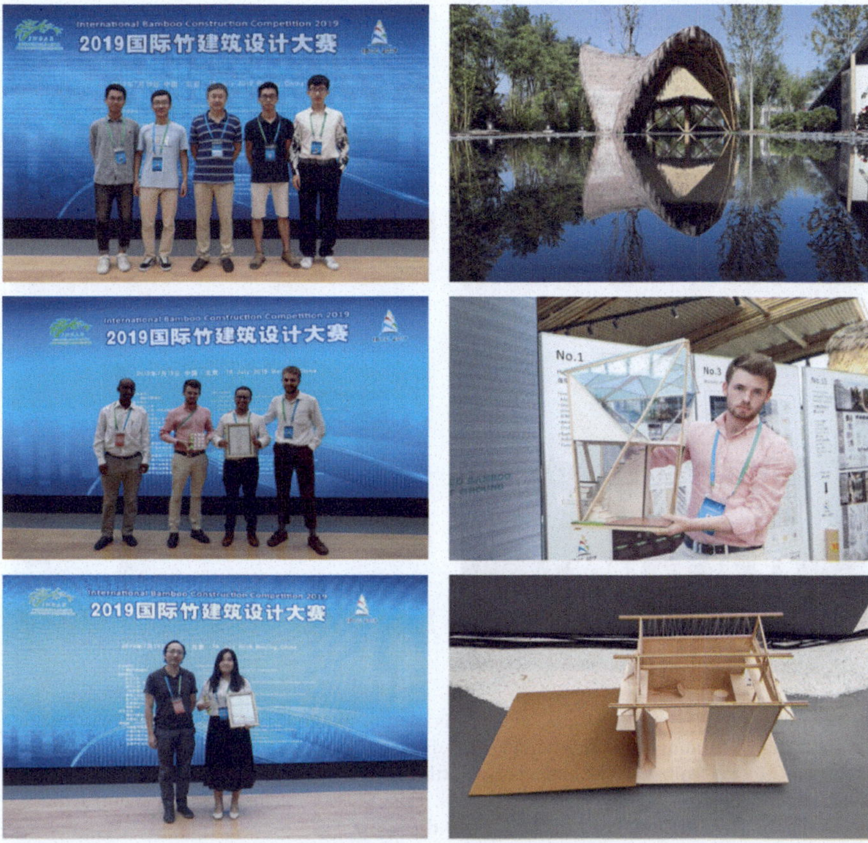

Fig. 10 Awarding Third Prize to Zhejiang University, University of Queensland and University of Chinese Academy of Sciences respectively

Fig. 11 Winner of the Public Vote: Beijing Jiaotong University

Fig. 12 Awarding Honorable Prizes to nine teams: Beijing Jiaotong University, Ethiopian Institute of Architecture, Building Construction and City Development, International Centre for Bamboo and Rattan, 3 teams from Xiamen University, North China University of Technology, 2 teams from Xi'an Jiaotong Liverpool University **a**, **b**, **c**; Ms. Hao Ying, Director of Membership Affairs is moderating the Awards Ceremony of IBCC 2019 **d**

Fig. 13 Trophy and certificate

Fig. 14 Ali Mchumo, Director General of INBAR awards certificates to all the onsite jury members of IBCC 2019 (From the left to the right: Prof. Zheng Xi, Prof. Yang Jun, Prof. Yan Xiao, Prof. Cui Kai, Ali Mchumo, Prof. Denamo Addissie, Prof. Gisela Loehlein, Ms. Liu Kewei, Prof. Zhao Chen, Prof. Zhang Li)

Fig. 15 Ali Mchumo, Director General of INBAR awards certificates to members of the Organizing Committee of IBCC 2019 (From the left to the right: Zhou Jianhui, Li Tao, Ms. Liu Kewei, Prof. Li Zhi, Ali Mchumo, Prof. Cristoforo Demartino, He Xiaozhou, Zhang Jinyuan, Wang Can, Zhang Mingming)

Fig. 16 Ali Mchumo, Director General of INBAR awards certificates to sponsor enterprises of IBCC 2019

Fig. 17 CGTN interview to the Winner Team and the Chair of IBCC 2019

Fig. 18 Public Vote of IBCC 2019 at INBAR Pavilion

a. Official book of IBCC 2019 "*Contemporary Bamboo Architecture in China*".

b. Logo of IBCC 2019 designed by students.

c. Bags.

d. Trophies for 16 prizes.

Fig. 19 Awards for IBCC 2019

Fig. 20 Organiser, Co-Organisers, Technical Support and Sponsors of IBCC 2019

Acknowledgements As the leading organizer, INBAR would like to extend our gratitude to the following co-organizers (Fig. 20) to support IBCC 2019: Zhejiang University, Zhejiang University-University of Illinois at Urbana Champaign Institute (ZJUI), School of Architecture of Tsinghua University, World Architecture, Federation of Hong Kong Industries Bamboo Industry Committee, Italian Bamboo Association, Base Bahay Foundation and Labirinto dealla Masone. We also thank to the great support from the jury committee, 23 global experts. Thanks to INBAR Construction Task Force and Arup to provide technical supports. Thanks to five sponsors to provide material or financial support: Ganzhou Sentai Bamboo Wood Company of China, Hongyazhuyuan Science and Technology Company of China, Zhujing Bamboo Company of China, Hangzhou Bamboo Technology Company, Beijing Sino-bamboo Culture and Technology Limited Company.

References

1. Liu, K.W., and O. Frith. 2013. An overview of global bamboo architecture: Trends and challenges. *World Architecture* 12: 27–34.
2. Jassen, J.J.A. 2000. *Designing and building with bamboo*. INBAR.

The manufacturer's authorised representative in the EU is Springer Nature Customer Service Centre GmbH, Europaplatz 3, 69115 Heidelberg, Germany. If you have any concerns regarding our products, please contact ProductSafety@springernature.com

Printed and bound by CPI Group (UK) Ltd, Croydon, CR0 4YY

01/10/2026

02226185-0005